International Political Economy Series

General Editor: **Timothy M. Shaw**, Professor of Commonwealth Governance and Development, and Director of the Institute of Commonwealth Studies, School of Advanced Study, University of London

Titles include:

Hans Abrahamsson
UNDERSTANDING WORLD ORDER AND STRUCTURAL CHANGE
Poverty, Conflict and the Global Arena

Andreas Bieler, Werner Bonefeld, Peter Burnham and Adam David Morton
GLOBAL RESTRUCTURING, STATE, CAPITAL AND LABOUR
Contesting Neo-Gramscian Perspectives

Morten Bøås, Marianne H. Marchand and Timothy M. Shaw (*editors*)
THE POLITICAL ECONOMY OF REGIONS AND REGIONALISMS

Sandra Braman (*editor*)
THE EMERGENT GLOBAL INFORMATION POLICY REGIME

Giorel Curran
21st CENTURY DISSENT
Anarchism, Anti-Globalization and Environmentalism

Martin Doornbos
INSTITUTIONALIZING DEVELOPMENT POLICIES AND RESOURCE
STRATEGIES IN EASTERN AFRICA AND INDIA
Developing Winners and Losers
GLOBAL FORCES AND STATE RESTRUCTURING
Dynamics of State Formation and Collapse

Bill Dunn
GLOBAL RESTRUCTURING AND THE POWER OF LABOUR

Myron J. Frankman
WORLD DEMOCRATIC FEDERALISM
Peace and Justice Indivisible

Marieke de Goede (*editor*)
INTERNATIONAL POLITICAL ECONOMY AND POSTSTRUCTURAL POLITICS

Richard Grant and John Rennie Short (*editors*)
GLOBALIZATION AND THE MARGINS

Graham Harrison (*editor*)
GLOBAL ENCOUNTERS
International Political Economy, Development and Globalization

Patrick Hayden and Chamsy el-Ojeili (*editors*)
CONFRONTING GLOBALIZATION
Humanity, Justice and the Renewal of Politics

Axel Hülsemeyer (*editor*)
GLOBALIZATION IN THE TWENTY–FIRST CENTURY
Convergence or Divergence?

Helge Hveem and Kristen Nordhaug (*editors*)
PUBLIC POLICY IN THE AGE OF GLOBALIZATION
Responses to Environmental and Economic Crises

Takashi Inoguchi
GLOBAL CHANGE
A Japanese Perspective

Kanishka Jayasuriya
STATECRAFT, WELFARE AND THE POLITICS OF INCLUSION

Dominic Kelly and Wyn Grant (*editors*)
THE POLITICS OF INTERNATIONAL TRADE IN THE 21st CENTURY
Actors, Issues and Regional Dynamics

Mathias Koenig-Archibugi and Michael Zürn (*editors*)
NEW MODES OF GOVERNANCE IN THE GLOBAL SYSTEM
Exploring Publicness, Delegation and Inclusiveness

Craig N. Murphy (*editor*)
EGALITARIAN POLITICS IN THE AGE OF GLOBALIZATION

George Myconos
THE GLOBALIZATION OF ORGANIZED LABOUR
1945–2004

John Nauright and Kimberly S. Schimmel (*editors*)
THE POLITICAL ECONOMY OF SPORT

Morten Ougaard
THE GLOBALIZATION OF POLITICS
Power, Social Forces and Governance

Richard Robison (*editor*)
THE NEO-LIBERAL REVOLUTION
Forging the Market State

Timothy J. Sinclair and Kenneth P. Thomas (*editors*)
STRUCTURE AND AGENCY IN INTERNATIONAL CAPITAL MOBILITY

Fredrik Söderbaum and Timothy M. Shaw (*editors*)
THEORIES OF NEW REGIONALISM

International Political Economy Series
Series Standing Order ISBN 0–333–71708–2 hardcover
Series Standing Order ISBN 0–333–71110–6 paperback
(*outside North America only*)

You can receive future titles in this series as they are published by placing a standing order. Please contact your bookseller or, in case of difficulty, write to us at the address below with your name and address, the title of the series and one of the ISBNs quoted above.

Customer Services Department, Macmillan Distribution Ltd, Houndmills, Basingstoke, Hampshire RG21 6XS, England

21st Century Dissent

Anarchism, Anti-Globalization and Environmentalism

Giorel Curran
Griffith University, Australia

palgrave
macmillan

First published 2006 by
PALGRAVE MACMILLAN
Houndmills, Basingstoke, Hampshire RG21 6XS and
175 Fifth Avenue, New York, N.Y. 10010
Companies and representatives throughout the world

PALGRAVE MACMILLAN is the global academic imprint of the Palgrave
Macmillan division of St. Martin's Press, LLC and of Palgrave Macmillan Ltd.
Macmillan® is a registered trademark in the United States, United Kingdom
and other countries. Palgrave is a registered trademark in the European
Union and other countries.

ISBN 13: 978–1–4039–4881–6   hardback
ISBN 10: 1–4039–4881–X   hardback

This book is printed on paper suitable for recycling and made from fully
managed and sustained forest sources.

A catalogue record for this book is available from the British Library.

Library of Congress Cataloging-in-Publication Data

Curran, Giorel.
   21st century dissent : anarchism, anti-globalization and environmentalism /
Giorel Curran.
      p. cm.
   Includes bibliographical references and index.
   ISBN 1–4039–4881–X (cloth)
      1. Dissenters. 2. Opposition (Political science) 3. Anarchism.
   4. Anti-globalization movement. 5. Environmentalism. I. Title.
   II. Title: Twenty-first century dissent.

   JC328.3.C87 2006
   303.48'4–dc22   2006047639

10   9   8   7   6   5   4   3   2   1
15   14   13   12   11   10   09   08   07   06

Printed and bound in Great Britain by
Antony Rowe Ltd, Chippenham and Eastbourne

For Tess

Contents

Preface

This book represents a coming together of several of my main research interests. My interest in environmentalism, particularly green political philosophy, goes back a long way, as does my interest in political theory and political movements in general. That quirky political philosophy anarchism had grabbed my attention right from the start, perhaps because it had been marginalized for so long, but probably because it had some very insightful political stories and ideas to tell. In the last few years it seemed that all these research areas came together in a very interesting form – the politics of anti-globalization. But my interest in the book's themes also goes beyond this. For those of us engrossed in the frequently chaotic and quickly changeable world of global politics, the early 21st century is already proving an immensely interesting, if increasingly worrisome, one. Not only are environmental risks reaching alarming levels, but so too are realignments in global power relations. Despite some significant improvements, 'old' problems of inequality and injustice remain. This is well illustrated in the increasingly inequitable distribution of environmental risks and the justice considerations they raise.

But all is by no means doom and gloom. As history consistently demonstrates, with injustice comes resistance, with appropriation comes counter-appropriation, and with hegemony comes counter-hegemony. This is precisely the undercurrent in contemporary oppositional politics that this book has sought to uncover. If this sounds utopian, we have to remind ourselves that the idea of enfranchising the 'great unwashed' was long considered a pipedream. One of human society's greatest achievements is undoubtedly democracy. Certainly, it took a long time for women and many Indigenous to be counted as full members of western democratic communities, and the struggle goes on in many other parts of the world. But democracy did make it on the agenda despite the concentrated resistance from many quarters of power, even if today its operation is at best frequently faulty and at worst an empty shell. The post-ideological anarchist impulse in contemporary dissent is a deeply democratic one. What is most optimistic about this oppositional current is its determination to continue pushing the democratic impetus by ensuring it incorporates the principles and practices of freedom, autonomy and equality. Utopian perhaps but neither unrealistic nor ahistorical.

In identifying this radical impulse for social change, it has been necessary to dissect, fragment and then reassemble it. This throws us into the murky territory of typology and classification – made particularly fraught when it is ideology that one is considering. Contemporary anarchism's ideological eclecticism, indeed intentional ideological capriciousness, has made neat classification a difficult task, even if this has not been the book's main intention. There is no definitive marker that neatly divides ideological from post-ideological anarchism, but there are strong indicators of the post-ideological temperament. These are the ones that we have sought to identify throughout, despite the messy and blurred residues that remain. There may be puzzlement over why some anarchisms do not appear and why some others have been included in the first place, and there is certain to be considerable grumbling among anarchists and post-ideological anarchists alike that their particular 'brand' has not been discussed, or discussed in an unsatisfactory manner. While I regret this, the fact remains that it is impossible to include all in the space constraints of one book. Difficult decisions also had to be made about what belonged, and what belonged where – as problematic as this task proved to be.

As usual, this work would not have been possible without the support of an assortment of colleagues, family and friends. Many have guided and supported me through the process of researching and writing this book. In various ways they have all helped strengthen this book. But I take all responsibility for its weaknesses which are mine alone and probably a result of not listening to their advice as carefully as I should have. Lists of names always raise the fear of accidental omission, for which I apologize in advance. I trust most of those I am grateful to know who they are in any case and I hope that I have already expressed my gratitude to them. But I wish to give a special thanks to Tess Curran, Jamie Curran, Yvonne Hartman, Keryn Hunter, Lesley Jenkins, Paula Cowan, Daniela Di Piramo, Robyn Hollander, Daniel Franks, Cathy Howlett, John Kane, Haig Patapan, John MacKenzie, Patrick Weller, and Elizabeth van Acker for the various support and advice they have provided throughout the life of this project. A special thanks also goes to Bruno Mezzalira who organized so many useful contacts for me and who helped make my research life while travelling so rich and interesting. I am also grateful to the Centre for Governance and Public Policy, Griffith University which has helped fund various stages of this project, and which has supported it throughout, and for a Griffith University Small Grant which enabled the timely conclusion of the project. I also wish to thank Simone Tosi from the University of Milan for organizing a very interesting and very useful seminar on the book's topic. Participants' feedback, suggestions and ideas have proved very useful, and for them I am grateful. More generally, many others were also involved with supporting and inspiring this project, and, once again, without naming them, I thank them all.

Giorel Curran
Brisbane

Introduction

Before the July 2005 G8 summit at Gleneagles, the Scottish media was awash with warnings of impending anarchist chaos. Determined to avoid another Genoa, the police force mounted one of its largest security operations in modern British history. They were particularly concerned with the rabble-rousing anarchists, suspected of plotting widespread disruption to the summit and elsewhere. In particular, security was trained on the 'notorious' Black Bloc who had clashed with police – and shopfronts – in past anti-globalization events. The Clandestine Insurgent Rebel Clown Army (CIRCA) – police harassment by tickling – and the anarchist People's Golfing Association (PGA) – police harassment by golfing – probably outnumbered Black Bloc type protesters. Yet police and media focus was set on the latter. Widespread reports of violent clashes between police and various anarchist groups outside the summit did eventually emerge. As it turned out, Bob Geldolf's 200,000 strong *Make Poverty History* march in Edinburgh snatched most of the attention. But all such news was swept aside in the wake of the London underground bombings at the beginning of the summit. In this light, anarchist posturing seemed even more petulant.

Anarchism has seldom had good press. Usually dismissed as either bomb throwing fanatics, eccentric utopians or idle scoundrels, anarchists have always struggled to have their political philosophy taken seriously. Unlike most of the other ideologies, anarchism's refusal to subscribe to vanguards, political parties or parliamentary politics denies it the traditional strategies for political success. Some historical examples have vindicated it, but this has not been enough to see it enjoy the authority of the major ideologies. Despite its relative marginalization as a political philosophy, anarchism has still exerted considerable influence in shaping the modern political landscape. More recently, a

1

particular mixture of socio-economic, cultural and political developments, and major technological advances, has created a political opportunity space for anarchism to both reassert and reinvent itself into its
influential 21st century incarnation. This has been achieved through the
medium of a largely anti-capitalist, anti-globalization and pro-green
global movement.

Despite anarchism's renewal, its contemporary influence has only
been cursorily acknowledged. The main objective of this book is thus
to explore the scope and tenor of this anarchist renewal, especially as
expressed in the radical ecology and anti-globalization movements.
It contends that the politics of globalization has propelled an invigorated anarchism into the heart of 21st century dissent. But the anarchism that it has unleashed is a considerably reconfigured one. The
term *post-ideological anarchism* is used to describe it. Post-ideological
anarchism informs the impulse, culture and organization of oppositional politics today. It refers to the looser and more flexible embrace
of anarchist ideas and strategies in the armoury of radical dissent. Post-
ideological anarchists are inspired by anarchism's principles and ideas,
drawing from them freely and openly to construct their own autonomous politics. They reject doctrinaire positions and sectarian politics, preferring to mix their anarchism with an eclectic assortment of
other political ideas and traditions. Post-ideological anarchism is also
primarily green.

Background

Anarchism's influence has evolved slowly, peaking and waning at different historical points. Refusing to be trampled under the weight of a
dominant Marxism, anarchists honed their alternative views as they
awaited what they saw as Marxism's inevitable implosion. The Spanish
anarcho-syndicalist experiments briefly showcased it, before Paris 1968
launched it as a serious contender in radical oppositional politics.
Anarchism then rode on the coat tails of the new social movements,
before poststructuralism and radical ecology sharpened its relevance to
contemporary politics. But it is in the early 21st century that anarchism
has come into its own, crystallizing in the anti-globalization politics of
the late 20th century.

Globalization has significantly transformed economics, politics and
culture across the globe. It is thus no surprise that the politics of globalization has framed and inspired anarchism's contemporary revival.
Globalization is, of course, a highly contentious and contested term,

described and understood very differently by a plethora of those affected by it. It encapsulates and describes important changes to global economic structures and the significant impact these changes have had on national and global economies, cultures and politics. The large numbers who feel passionately about globalization tend to identify as either its supporters or opponents. But it is globalization's opponents that have been considerably more vocal, and who have articulated their opposition in more visible, expressive and combative ways. This helps explain the high visibility of the anti-globalization movement – or more aptly the global justice movement or 'movement of movements' – with its diversity of participants and forms of dissent. The anti-globalization movement represents a highly visible and active constellation of resistance against the ills of globalization, especially a globalization underpinned by neo-liberal values. It is in this antagonism to neo-liberal globalization that anarchist ideas have found much resonance, in turn helping drive the angst of the anti-globalization movement.

Globalization is an important explanation, but the factors driving this quasi anarcho-renaissance are in reality complex and varied, and precede the 'formal' advent of globalization. Several main factors have helped launch modern anarchism. First, while anarchism has a long historical pedigree, the crises of the communist experiment, both pre- and post-1989, and the consequent fracturing of the left, reawakened an interest in anarchist thought. The contest between anarchism and Marxism goes back a long way, but the fracturing of the socialist alternative has opportunely re-positioned contemporary anarchism. While Fukuyama's (1992) 'end of history' claim is problematic in a number of ways – not least in the claims it makes for a triumphant liberalism – it does correctly identify a significant destabilization of the major political alternative – communism, and the considerable fragmentation of the left that resulted. Disillusioned and disappointed with the problems of communism, some on the left readily embraced an anarchist analysis that had consistently cautioned against the authoritarian and vanguardist trappings of socialism. This disillusionment was reinforced by the vigour of capitalism's latest stage – neo-liberal globalization – and the seeming impotence of the 'old' left in its wake.

Communism's crises have thus gone a considerable way towards ideologically validating anarchism's antipathy to it. And when Soviet communism collapsed in 1989 this vindication was seemingly complete. While anarchists and Marxists have long shared their opposition to capitalism and the socio-political relations it generates, anarchists have long contended that the Marxist conceptualization of power was short-sighted. It was in the failure to locate hierarchy and the central-ization of authority as the key drivers of oppression, that the anarchists foresaw the crumbling of socialism. Bakunin had rebuked Marx and his followers long ago as 'worshippers of the power of the State' and as 'the prophets of political and social discipline, champions of the social order built from the top down' (in Marshall 1993, 303). The ruthless centralization of power exhibited in the USSR was to render prophetic the predictions of Bakunin and like-minded anarchists. Vindication lay in the anarchists' identification of an underpinning authoritarianism as Marxism's major blind spot. This is not to say that this vindication led to a widespread embrace of anarchism; far from it. But it did enlarge the political opportunity space within radical politics that anarchism was able to comfortably fill. With the advent of neo-liberal globalization and communism's retreat, anarchism was well placed to rally a disenchanted left in considerable disarray.

The New Left had already paved the way for this enchantment with anarchism. New Left analyses, and the discourses of postmodernism and poststructuralism, resonated anarchist sensibilities. In challenging the Marxist orthodoxy – its historical materialism, economic determin-ism and class politics – and in promoting an expanded account of the practice of domination, the New Left won itself numerous oppositional friends, including anarchists. While still acknowledging the structural underpinnings of capitalism, the New Left was equally interested in the cultural, psychological and aesthetic patterns of domination, patterns a narrowly-focused Marxism neglected to address. In broadening the conceptualization of domination, the New Left helped identify a more extensive range of 'disciplinary' practices that together maintained oppression. They hence challenged the limitations of Marx's economic determinism and working class praxis as the motor of social change. The New Left also influenced and informed the budding radical eco-logy movement. Drawing from some New Left analyses, these radical ecologists went further, condemning the androcentric, technocentric and anthropocentric underpinnings of capitalism and industrialism as well as of Marxism. The counter-culture of the 1960s embraced this expanded critique since it represented not only a liberation from the stranglehold of 'old' leftism, but also better accommodated their specific grievances. The ensuing focus on increased autonomy and life-style alternatives helped launch the new social movements of the 1960s and 1970s. With them came a widespread dissemination of anar-chist ideas.

The late 1960s is thus frequently marked out as a historical turning point for oppositional politics. The poststructuralist, post-Marxist and anarchical impulses that animated the Paris revolt in 1968 were underpinned by a distaste for modernism and the Enlightenment legacy that had promised much but delivered little. Feminists, the colonized, people of colour, queer activists and advocates for the rights of nature specifically denounced the exclusive politics of both the traditional left and right, arguing instead for an inclusive practice and 'politics of difference'. Difference was celebrated in a variety of cultural expressions: a spirit of anti-authoritarianism, freer sexual politics, a celebration of different life-styles and dress codes, and a variety of Do-it-Yourself direct action politics, including political 'carnival', 'spectacle' and early forms of 'culture jamming'. With modernism increasingly challenged, Paris 1968 became 'the cultural and political harbinger of the subsequent turn to postmodernism' (Harvey 1989, 38), signifying the dawning of a new politics.

Postmodernism and poststructuralism rode, as well as drove, this wave of new politics, albeit taking it in different directions. Anarchism was heartened to see some of its ideas accommodated in the new discourses and the new politics, but it was at the same time challenged by them. While many anarchists were sceptical of what they saw as postmodernism's *a*political nature, many others embraced the insights of poststructuralism, using them to reshape and revitalize anarchist thought itself. Anarcho-communists and other collectivist anarchists, after all, borrowed considerably from an 'unreconstructed' Marxism hampered by structuralist limitations. The new anarchism that emerged – a broad and eclectic collection of new anarchist schools, theories and ideas – drove anarchism's own internal renewal. Through addressing its own modernist and ideological limitations, anarchism sought to better position itself to take advantage of the refashioning of radical politics. This refashioning included an embrace of radical greens who were also beginning to identify in modernism's instrumentalist logic, the tools used to dominate nature.

Together these political and philosophical developments represented a horizon of new opportunities for anarchists – a relatively open market for political alternatives in which they could showcase their wares. Post-1989 in particular had birthed a transformed political landscape. Many of these anarchists now believed that liberal capitalism has not yet confronted a truly formidable ideological adversary such as contemporary anarchism set out to be. But unlike an allegedly stolid socialism, anarchism would be a tricky, savvy and footloose adversary.

It would be 'remade' and it would be stronger. As the contemporary anarchist Bob Black contends, 'anarchists are [now] at a turning point. For the first time in history, they are the only revolutionary current' (Black 1997, 140). In making this claim, Black may have been overstating his case. But he makes an important point. He identifies an open political space through which to (re)launch the anarchist imaginary. The politics of the past few decades had propelled the anarchist impulse, but the emergence of virulent anti-globalization represented the opportunity to drive it home.

A post-ideological anarchism for the 21st century

Anarchism has embraced the reconfigured ideological landscape of the early 21st century and made it its own. Radicals disillusioned with the capacity of traditional oppositional ideologies to challenge capitalism and neo-liberalism, find its analysis increasingly appealing. These radicals observe not only the ravages of neo-liberal globalization, and socialism's weakness in stemming them, but also an environmental ruin that critically threatens both people and planet. They find particularly disturbing a new century in which one major ideology, liberalism, has morphed into an even more damaging incarnation – neo-liberalism; and the other, socialism, has proved increasingly ineffective in challenging it. As Kinna (2005, 21) points out, one of contemporary anarchism's 'striking features' is its 'conviction that political and cultural conditions have altered so radically in the course of the twentieth century that the traditional schools of thought ... have become outmoded'. This has catapulted anarchism's 'culture and forms of organization ... to the forefront rather than the margins of a transnational social movement' (Milstein 2004). In short, the contemporary combination of an anti-capitalist surge fuelled by globalization; the concerns of ecology; the left's political reflection in the face of many setbacks; and the availability of sophisticated technologies, has significantly reanimated anarchism (see Curran 2004a). But this reanimated anarchism is a differently configured one.

This book uses the term post-ideological anarchism to capture this reconfigured anarchism. Influenced by developments we described above, a post-ideological anarchism is conditioning the spirit and practice of radical dissent today. It is an anarchism freed from ideological conformity and one that borrows openly from a panorama of ideas and traditions. There remain, of course, many ideological anarchists who still participate as proud anarchists in oppositional protest. Some of the new anarchist schools, along with the old, continue to assume highly ideological positions. But, more importantly, there is the looser and widespread embrace of anarchist ideas and strategies within the armoury of radical oppositional politics. Here different forms of dissent are largely *inspired* by the ideas and animating principles of anarchism. In a post-ideological spirit, these radicals feel at liberty to draw from the force of anarchism's ideas flexibly and non-doctrinally, without necessarily identifying as anarchist. Instead these 'small-a anarchists' pull and pluck from the ethical force of anarchism to remake it in a manner that suits their own autonomous objectives (Neal 1997). It is this anarchist *impulse* percolating through oppositional politics today, that represents a primary way in which anarchism is influencing contemporary dissent.

Anarchism's core values remain autonomy, liberty, anti-statism and anti-authoritarianism. It continues to see hierarchy, authoritarianism and the centralization of decision making power, both within the state and elsewhere, as inimical to the achievement of those values. And commitment to a correspondence between means and ends still underpins anarchism's strategic heart. As a libertarian and anti-authoritarian political philosophy, anarchism has an overriding allegiance to the principles of radical democracy – preferably direct, certainly participatory and always transparent and inclusive. But to this list of core values has now been added a *green* one. Anarchism, particularly new anarchism, has enthusiastically embraced the claims of radical ecology that environmental degradation signifies the enhanced destructive power of industrialism and/or capitalism. Now most modern anarchists have incorporated, either centrally or more peripherally, the claims of ecologism, agreeing that the will to power degrades both people and nature. But in the 21st century these core values, and the strategies to achieve them, are increasingly interpreted and assembled differently. This diverse assemblage, accommodated in much of the new anarchism, draws from the classical greats, and other traditions, in a looser and less doctrinaire fashion – a development that many new radicals find appealing.

Other commentators have made similar observations, and we quickly review some of them below. While we build on these observations, our conceptualization of post-ideological anarchism goes further. We identify and probe in considerable detail the diverse elements that constitute the mosaic of post-ideological anarchism, before tracing it in a number of illustrative case studies. We also insert a decidedly green ethos into its centre.

Neal (1997) goes closest to prefiguring important aspects of our post-ideological anarchism. He distinguishes between what he calls small 'a' and capital 'A' anarchism, the former denoting a less ideological strand than the latter. More specifically, he conceptualizes a capitalized Anarchism as an ideology and the lower case anarchism as a methodology. As an ideology anarchism becomes 'a set of rules and conventions to which you must abide' while as a methodology it is 'a way of acting, or a historical tendency against illegitimate authority' (1997). He observes that:

> Sadly, what we have today are a plethora of Anarchists – ideologues – who focus endlessly on their dogma instead of organizing solidarity among workers. That accounts for the dismal state of the movement today, dominated by elites and factions, cliques and cadres ... Methodology is far more open – there is that which works, and that which doesn't, and degrees between those points. If one strategy doesn't work, you adjust until you get something that does work (1997).

For Neal, a dogmatic Anarchism violates the true spirit of anarchism. He believes that anarchist organization cannot be proscribed, but should arise spontaneously from the autonomous community that conceives it. Nor can an 'indoctrinated people' be a free people. If the capacity to decide principles and strategies are denied them, such people are both not free and not anarchist. But writing in 1997, Neal may have been heartened by the spirited defence of his small 'a' anarchism in the subsequent politics of anti-globalization.

Graeber (2002, 72) utilizes Neal's distinction to help explain the influence of anarchism today, and agrees with him that even in 2002 there are many capital-A anarchist groups. Importantly, however, he believes that the small-a anarchists – those non-card carrying radicals in the anti-globalization movement inspired by the principles and moral force of anarchism – 'are the real locus of historical dynamism right now'. While he still contends that anarchism has an ideology, it is a non-sectarian and deeply democratic one:

> A constant complaint about the globalisation movement in the progressive press is that, while tactically brilliant, it lacks any central theme or coherent ideology ... [But] this is a movement about re-inventing democracy. It is not opposed to organization. It is about creating new forms of organization. It is not lacking in ideology. These new forms of organization *are* its ideology (Graeber 2002, 70)

Epstein (2001) too notes the attraction of looser, non-doctrinaire anarchist positions for the new generation of young radicals not formally schooled, or even interested, in the radical tradition. She contends that while anarchism has always attracted many young radicals, those in the anti-globalization movement today are not necessarily interested in old dead anarchists, or in anarchism as a body of theory. But they are inspired by many of its principles and impelled by its vision. Indeed, for younger radicals:

> [A]narchism means a decentralised organisational structure, based on affinity groups that work together on an ad hoc basis, and decision-making by consensus. It also means egalitarianism; opposition to all hierarchies; suspicion of authority, especially that of the state; and commitment to living according to one's values (Epstein 2001, 61).

She utilizes a useful way of understanding and conceptualizing contemporary anarchism that echoes our conceptualization of post-ideological anarchism. In determining anarchism's influence she distinguishes between anarchism *per se* and anarchist sensibilities, between those who identify with anarchism as a tradition and ideology and those who simply identify with its spirit and the force of its ideas. In short, she draws a distinction between 'ideological' anarchism and an inspirational anarchism that resonates post-ideological anarchism. Writing in the late 1990s, Purkis and Bowen (1997, 3) identify a similar phenomenon, arguing that the 'terrains of theory and action have changed' so that 'now there are generations of activists operating in many fields of protest for whom the works of Kropotkin, Malatesta and Bakunin are as distant ... as ... Charles Dickens'. In their more recent work, they note the considerable change that anarchism has undergone, especially in its broader conceptualization of power (Purkis & Bowen 2004).

In a similar vein, new anarchist theorists themselves highlight a comparable phenomenon, both as it influences internal theory and external politics. 'Postanarchist' theorists highlight similar developments. Adams (2004) for example, distinguishes between those who identify with anarchism as an 'ideological tradition' and those who identify with its 'general spirit'. He contends that postanarchism's post-ideological character is reflected in the fact that 'it is not an "ism"' nor 'another set of ideologies, doctrines or beliefs' that together act as a 'bounded totality' to which one conforms (2004). Rather than subscribing to 'ideological anarchisms such as anarchist-syndicalism, anarchist-communism, and anarchist-platformism' postanarchism manifests today:

... not only in abstract radical theory but also in the living practice of such [anti-globalisation] groups as the No Border movements, People's Global Action, the Zapatistas, the Autonomen and other such groups that while clearly 'antiauthoritarian' in orientation, do not explicitly identify with anarchism as an ideological tradition so much as they identify with its general spirit in their own unique and varying contexts, which are typically informed by a wide array of both contemporary and classical radical thinkers (Adams 2004).

A fellow postanarchist concurs:

[There] are the equally if not more important, growing numbers of people who just feel dissatisfied with 'all' ideologies in general, yet who can also sense the profound resonance a nondoctrinaire anti-authoritarian analysis has within contemporary social movements (Bey in Adams 2004).

The new anarchists Bob Black, and Hakim Bey after him, talk about 'type 3 anarchism'. This is a type of 'radically non-ideological' anarchism that is 'neither Individualist nor Collectivist but in a sense both at once' (Bey 1991). For Black (n.d.), while type 3 anarchism resists categorization, he still distinguishes it from the other two types. Type 1 refers to anarcho-leftism and type 2 to anarcho-capitalists, even though he is quick to dismiss them as unrepresentative of the anarchist tradition. But it is type 3 that identifies the contemporary anarchist moment:

The worldwide, irreversible, and long-overdue decline of the left precipitated the current crisis among anarchists... Anarchists are having an identity crisis. Are they still, or are they only, the left wing of the left wing? Or are they something more or even something else? Anarchists have always done much more for the rest of the left than the rest of the left has ever done for them. Any anarchist debt to the left has long since been paid in full, and then some. Now, finally, the anarchists are free to be themselves (Black n.d.).

Black's type 3 anarchists are thus free to draw from Situationism or syndicalism, Marxism or Islamism, feminism or Christianity and a plethora of other, even contradictory, influences. The key to type 3 is its political openness, diversity, non-sectarianism and autonomy.

Finally, if in a somewhat different vein, Day (2004) identifies in contemporary radical politics a shift from the 'hegemony of hegemony' to 'non-hegemonic forms of radical social change'. By this he means that if the goal of social transformation is to be achieved, radical change has to be less hierarchical in its spirit and organization. He locates in the anti-globalization movement just such an awareness, one driven by what he calls a 'logic of affinity'. This logic resembles Hardt and Negri's (2000; 2004) constituent power of the multitude, but is strengthened by the utilization of anarchist insights. A logic of affinity built on anarchist theory and practice is discernible in the anti-globalization movement today. Day (2004, 740) articulates the key elements of this logic:

> ... a desire to create alternatives to state and corporate forms of social organisation, working 'alongside' the existing institutions; proceeding in this via disengagement and reconstruction rather than reform or revolution; with the end of creating not a new knowable totality (counter-hegemony), but of enabling experiments and the emergence of new forms of subjectivity; and finally, focusing on relations between these subjects, in the name of inventing new forms of community.

In short, Day identifies – albeit on the basis of supporting a different argument – some of the ingredients of post-ideological anarchism. He situates a non-hegemonic anarchist impulse, akin to our post-ideological one, at the centre of radical social change. In the process he notes, following Graeber, that 'if anarchist-influenced groups look disorganized' this is because they practice a non-hegemonic form of organization that the traditional left, still locked into hegemonic political practices, ridicule (Day 2004, 741).

The above examples help illuminate how post-ideological anarchism separates itself from traditional, 'ideological' anarchism as well as traditional left politics as a whole. Within the tradition of ideological anarchism can be located specific anarchist schools that assume sectarian and doctrinaire positions: the capital A anarchists. Within the AGM, we also find activists who are members of specific anarchist schools and who practice their oppositional politics accordingly. However, the contemporary face of anarchism is best represented in terms of key anarchist sensibilities that have penetrated the modern protest lexicon and helped shape visions of socio-political alternatives. Here anarchism is not necessarily swallowed 'holus bolas' but its considerable narrative force informs and inspires much of the AGM and the arena of radical politics as a whole, making it the 'locus of dynamism' that Graeber identifies. Many dissenters in the AGM do not self-consciously identify as anarchist, even if they incorporate key elements of anarchist organization and wear its principles proudly. Importantly, removal from the demands of ideological purity permits a new eye to be cast over the offering of other ideologies, and of the novel incorporation of some of these strands into a reworked post-ideological anarchism. The Zapatistas exemplify this autonomous, anarchical praxis well – but without needing or wishing to identify as anarchist, or socialist for that matter.

Not unexpectedly in a politics that champions diversity, there are significant tensions within post-ideological anarchism. Most of these tensions are long-standing but emerge in different ways in new anarchist thought. They also mirror some of the theoretical and strategic dilemmas that dog radical politics generally. The renewed debates about individual *versus* social anarchism raise the perennial dilemma of oppositional agency – of whether an individualist or collectivist political strategy is more likely to realize set objectives. The issue of technology is also central. Views on its usefulness ranges from anarcho-primitivist Luddites who reject it altogether, to its embrace as a key organizational and political tool by the anti-globalization movement. Despite the influence of postmodernism and poststructuralism on contemporary anarchism, there still remains considerable caution on the value of its offerings. Finally, the issue of violence and the 'directness' of direct action, remains central. Anarchism has long been associated with violence and chaos. The 'propaganda in the deed' tactic has helped generate the association between anarchism and violence, an association now fuelled by the property violence of the Black Bloc and the Earth Liberation Front. Nonetheless, the majority of anarchists, particularly those within the anti-globalization movement, neither practice nor sanction such tactics. We observe these tensions in the discussion of our case studies.

To summarize, this book contends that a post-ideological green anarchism is increasingly influencing the impulse, culture and organization of 21st century dissent. It is an anarchism that rejects the stranglehold of ideology by discarding doctrinal purities and vanguardist politics. While still adhering to some of the insights of the 'old greats,' post-ideological anarchism reveals a hybridization of a number of different influences and traditions. Its 'post-left' character has attracted a significant following, especially among young radicals for whom the old left and traditional ideology is a dim speck on the political horizon.

Small 'a' anarchists inspired by anarchist values are more likely to draw from the writings of some of the new anarchists than the old. Complex philosophical tomes attract very few of them, but they will avail themselves of briefer reads on the internet, where most new anarchist material is 'copy-lefted' and accessible. They are largely drawn to anarchism's spirit and the ideological freedom its staunchly autonomous, individualist ethos permits them. These radicals are particularly compelled by anarchism's network and affinity group structure, a structure facilitated by the new technology, and widely practiced within anti-globalization protest. In short, anarchism – post-ideological and green – has emerged as a viable force in the oppositional politics of the 21st century.

The chapters

The book is divided into two main parts. The first – Theorizing Contemporary Anarchism – explores the theoretical influences and political developments that have stimulated the shape of contemporary anarchism and its post-ideological expressions. We have briefly identified some of these themes in this introduction. While neat classification of a consciously fluid, flexible and eclectic 'position' clearly presents its own organizational difficulties, in the second part we nonetheless utilize a number of case studies to illustrate how this influence is expressed and enacted. The cases in this second part – Practicing Contemporary Anarchism – focus primarily on groups who participate in or support the spirit of the anti-globalization movement.

The first chapter, *Anarchism Old and New*, traces the development of anarchist thought from its classical beginnings through to the newer anarchist schools and ideas. It identifies anarchism's core values and how they their interpretation and application have altered over time. The chapter sacrifices a fuller discussion of classical anarchism to focus on some of the developments within new anarchism – developments that have thus far received relatively less attention. New anarchism remains indebted to classical anarchism, but it also draws from a broader range of sources. In undertaking a considerable remodelling, new anarchism has influenced and informed post-ideological anarchism. Not unexpectedly the new schools, along with the old, do not necessarily agree with each another and are often fragmentary and only partially developed. We devote considerable attention to the tension between individual and social anarchism, particularly since this tension re-emerges in the politics of the new anarchism.

Chapter Two – *Movements of Anti-Globalization* – explores how and why these movements were key to propelling post-ideological anarchism into the heart of radical dissent. While 'movement of movements' or 'global justice movement' describe it better, we continue to use the term anti-globalization to help situate radical dissent in the broader politics of globalization that underpin today's political landscape. This chapter identifies the anti-globalization movement's (AGM) key features before distinguishing it from the new social movements that preceded it. In the process it notes the considerable influence the green movement has had on shaping the AGM. Signalling its focus on global justice, the AGM's gaze was trained on the link between globalization and inequality and between the trashing of ecology and society. Neo-liberal globalization was consequently identified as the 'enemy' and as the direct source of discontent. The movement's anarchical impulse and organizational structure are illustrated throughout, finishing with an overview of the World Social Forum's role in cohering the movement.

The role that technology plays in animating and operationalizing the AGM, and contemporary oppositional politics as a whole, forms the basis of this next chapter. Chapter Three – *Technologies of Dissent* – examines how new technology enables an anarchical style of networked protest, one that is decentralized, acephalous and non-hierarchical. The chapter also uncovers some interesting synergies between anarchical sensibilities and 'anarchical' technologies such as the internet. But radicals still view technology as a two-edged sword: it is on the one hand a social control tool constructed and wielded by the powerful; on the other, oppositional politics has counter-appropriated it for radical purposes. This is a main paradox of the anti-globalization movement. Its opposition to globalization is reinforced by the capacity to spread its dissent globally through the very technologies of globalization.

Chapter Four – *Ecology and Anarchy* – investigates the emergence of radical ecology and its links to anarchism. Anarchism's greening has been underway for some time, with the two discourses drawing from and informing each other. Utilizing critical theories sympathetic to anarchism, the green movement – disillusioned and disappointed with the failures of Marxism to account for industrialism's devastating impact on nature – readily embraced an anarchism that expanded the discourse of domination beyond class. Many radical greens were heartened by anarchism's identification of hierarchy as central to the operation of domination. At the same time, anarchists recognized in radical ecology issues and values that went to the philosophical and political heart of anarchism. This chapter explores the various synergies between anarchism and ecology in a range of radical ecology discourses.

The Zapatista uprising in Mexico in 1994 captured the world's protest imagination. This event represented an important turning point in late 20th and early 21st dissent. the Zapatistas' direct and dramatic articulation of globalization's ills helped inspire global dissent against neoliberal globalization. In doing so, they were instrumental in showcasing a post-ideological 'tactical template' that helped define the political character of 21st century dissent. This fifth chapter – *The Politics of Zapatismo* – discusses the rise of the Zapatista movement and the development of its political philosophy – Zapatismo. In particular, it identifies the elements that give Zapatismo its distinctiveness and resonance as a global politics of dissent. The Zapatistas are not anarchists, nor socialists – indeed they resist such labels. But they draw from anarchism flexibly and non-doctrinally to assemble their very specific form of *autonomous* politics. In seeking to 'exercise' power rather than 'take' it, Zapatismo is clear in its refusal of vanguardist and sectarian politics. In doing so, it constructs a post-ideological politics that is autonomous, flexible and reflexive – and, for many, inspirational.

Chapter Six – *Greening Anarchy: Social Ecology* – discusses one of the most developed and influential green anarchist schools – Murray Bookchin's social ecology. Social ecology is an innovative and eclectic social theory that explores the implications of domination and hierarchy for both society and the environment. It utilizes an extensive range of historical and anthropological data in locating the social origins of ecological crisis. While social ecology is staunchly anarchist and decidedly green, Bookchin is less comfortable in a post-ideological environment than other radicals. Indeed, he caustically rebuffs contemporary anarchism's post-ideological character – referring to it perjoratively as 'lifestyle anarchism'. The chapter explores the breadth of the social ecology discourse before considering its place in a post-ideological anarchist politics.

Reclaim the Streets – the subject and title of Chapter Seven – refers to both a tactic of dissent as well to the groups who organize under its banner. Originating in the UK, it has influenced the creation of national chapters throughout the world. Strongly affiliated with Earth First!, Reclaim the Streets (RTS) situates autonomous direct action at the heart of their politics. Highly symbolic and highly visible, RTS seek to re-appropriate public space from the enclosure of globalization. Its commitment to spontaneity, autonomy and diversity demonstrates its anarchical temperament. So too does its championing of an organizational form that practices autonomy through the (temporary) occupation of space outside state control. Drawing post-ideologically and flexibly from a range of political sources, particularly Situationism, RTS counter-appropriates dominant cultural tools for subversive – and playful – purposes.

Often viewed as direct action exemplars, Earth First! has been at the forefront of radical ecology actions across the globe. This final chapter, *Earth First!*, explores how in steadfastly refusing to negotiate with the state, Earth First! actions aim squarely at the perpetrators of ecological damage, targeting them explicitly through a range of direct actions, or through the reclamation of space in such actions as Reclaim the Streets or the anti-roads movement. The organizational principles of Earth First! testify to its anarchist credentials: it is non-hierarchical, encourages membership diversity, rotates coordinator roles, and ensures its chapters are autonomous and independent. It contains a diverse mix, including many post-ideological anarchists practicing a flexible and individualized direct action politics under its broad banner. The Earth Liberation Front (variously accepted or rejected as an offshoot of Earth First!) represents the extreme end of a direct action politics.

Overall, this book discusses anarchism's new form and its participation in the politics of contemporary dissent. But in identifying the character of radical politics in the 21ˢᵗ century it raises broader implications for global politics as a whole. We speculate on some of these implications in the broader conclusions drawn in the book's closing chapter.

Part I

Theorizing Contemporary Anarchism

1
Anarchism Old and New

Anarchism is complex, multifaceted and fragmented – like most ideologies. Its history is a rich and diverse one, and it boasts many theorists whose contributions to political philosophy stand proudly alongside the weight of the liberal and Marxist ones. It contains a core set of principles to which most anarchists subscribe, but these principles are broadly interpreted and diversely applied within the panorama of ideas and schools that it accommodates. This breadth includes an anarchism of the left and an anarchism of the right, even if much of the left is loathe to embrace the right's anarcho-capitalism into its fold. Anarchism has been described as a fertile ideology with a 'broad back' (Guerin 1970), one that contains an eclectic mix of theorists, activists, ideas and strategies. But in refusing to adopt traditional routes of political transformation, and in its long and unflinching opposition to the state and its institutions, it has long struggled to be taken seriously as a 'practical' political ideology. Often dismissed as hopelessly utopian, the rich intellectual and philosophical offerings of anarchism gathered considerably more dust than most of the other traditions. But many see in the anti-capitalist politics of the early 21st century the dusting off of an anarchism whose relevance resonates with an increasing number of radicals, and whose principles are helping to shape contemporary dissent.

This chapter reviews anarchism's core values, its historical development and its contemporary expressions. One chapter cannot do justice to the breadth of anarchist thought, and fuller discussions of classical anarchism is left to the many excellent books that do just this. This allows us to devote more space to exploring the more recent developments within 'new' anarchism, as well as the tensions between social and individual anarchism – tensions that have recently reemerged in the new anarchism. The chapter is thus divided into two main sections: old and new anarchism. While a useful organizational devise, such a distinction can obscure the significant continuities between the two, and the fact that 'new' anarchism draws from 'old' or classical anarchism in direct and important ways. But there are significant theoretical developments within new anarchism that justify this distinction and our attention. These developments also help inform contemporary anarchism and its post-ideological face.

To anarchists, the values of liberty and autonomy are everything, and they staunchly resist all attempts to trample them. Since the state is the acme of authority, as well as the institutionalized site of domination and control, anarchists defy it passionately. The anarchist core is thus an anti-authoritarianism that opposes all forms of mediated, centralized and hierarchical power. Since they see the state as representing and working for the interests of the powerful, and since it uses its considerable institutional infrastructure to deny individual integrity, there can be no accommodation with it. The shared anarchist objection to the state as a core principle is not as straightforward as it may first appear, however. Anti-statism differs considerably in degree and emphasis among anarchism's diverse mix. There is much disagreement about the kind of relationship individuals and communities should have with the state, especially before and during the process of its dismantling, and disagreement on the preferred means for its utilization, destabilization or outright overthrow.

Rudolph Rocker expands on the set of core principles to which most anarchists, with the exception of anarcho-capitalists, would subscribe:

> ... personal and social freedom is conceivable only on the basis of equal economic advantages for everybody ... the war against capitalism must be at the same time a war against all institutions of political power, for in history economic exploitation has always gone hand in hand with political and social oppression. The exploitation of man by man and the domination of man over man are inseparable, and each is the condition of the other (Rocker 1938, 17–18).

And certainly most anarchists would empathize with Proudhon's classical and strident reproach of the state:

> To be governed is to be kept in sight, inspected, spied upon, directed, law-driven, numbered, enrolled, indoctrinated, preached at, controlled, estimated, valued, censured, commanded, by creatures who

have neither the right, nor the wisdom, nor the virtue to do so ... to be governed is to be at every operation, at every transaction, noted, registered, enrolled, taxed, stamped, measured, numbered, assessed, licensed, authorised, admonished, forbidden, corrected, punished ... That is government. (Proudhon 1989, 294)

Of course it is in the strategies for overthrowing such oppression that considerable disagreement emerges, with different anarchist groups championing their preferred tactics against a range of other comers. While it is not the intention of this chapter to step into classificatory complexities, some disentangling is necessary. We begin our discussion with classical or 'old' anarchism before exploring the intricacies of its new face.

Old anarchism: classical debates

A prominent contemporary anarchist, Noam Chomsky, names his anarchism as one steeped in the classical tradition. He refers to some of the great social or collectivist anarchists – particularly the communitarian 'greats': Bakunin, Proudhon, Kropotkin – as major influences on his anarchist inclination. This approach invokes one of the earliest anarchist typologies: the 'classical' anarchist theoreticians on the one hand, and 'all the rest' on the other (see Kinna 2005, 11). Over a hundred years ago, Eltzbacher (1908) identified seven key anarchists, distinguished by the central theoretical contributions they made to the study of anarchism. Most of these largely 19th century greats – Pierre Proudhon, William Godwin, Max Stirner, Michael Bakunin, Peter Kropotkin, Benjamin Tucker and Leo Tolstoy – continue to be nominated today as anarchist celebrities (Woodcock 1963; Ritter 1980; Miller 1984; Marshall 1993; Kinna 2005, 10–15).

Anarchism draws from two main and seemingly oppositional ideological traditions: liberalism and Marxism. Described by an anti-globalization activist as 'liberalism on steroids', anarchism champions the individualism and autonomy that liberalism also promotes, albeit rendered differently. Some anarchists attach this love of liberty to capitalism, seeing in the 'free' market an ideal way of enhancing and rewarding individual endeavour. Anarcho-capitalism, most prominent in the United States, has produced such luminaries as Benjamin Tucker and more recently, Ayn Rand and Murray Rothbard. Heated dispute remains over whether anarcho-capitalists should be accepted into the anarchist fold in the first place. All agree however that the larger branch of left, particularly social, anarchism draws much from the Marxist tradition, even if it eschews large analytical chunks, especially its revolutionary strategy. The classical luminary Kropotkin calls his anarchism communist anarchism, and anarcho-syndicalism has been highly influential in the working class politics of the 19ᵗʰ and 20ᵗʰ centuries. The dual influences of liberalism and socialism have generated the term libertarian socialism as more apt a description of this form of anarchism. Chomsky proudly identifies his form of anarchism as libertarian socialism and in so doing, highlights anarchism's immersion in the 'best' of enlightenment and classical liberal thought. This reveals not only the breadth of the anarchist imaginary but also the eclecticism of its influences.

In agreeing that his anarchism approximates a 'socialism with freedom', Chomsky's libertarian socialism captures a form of anarchism that quite clearly distinguished itself from the individualist extremes of a Max Stirner, as well as the anarcho-capitalists of a proud right tradition. The idea of libertarian socialism also reveals the dual attachment to individual freedom and collectivism that most anarchists subscribe to. When asked in an interview whether he found it ironic that during the Spanish revolution an anarchism renowned for its advocacy of individual freedom ended up enacting an organized and democratic collectivism, Chomsky replies that:

> The tendencies in anarchism that I've always found most persuasive seek a highly organised society, integrating many different kinds of structures (workplace, community, and manifold other forms of voluntary association), but controlled by participants, not by those in a position to give orders (except, again, when authority can be justified, as is sometimes the case, in specific contingencies) (Doyle 1995).

Like many anarcho-socialists before him, however, Chomsky distinguishes his anarcho-socialism from a Marxist-Leninism set on capturing state power. He invokes Bakunin's historical prediction that the revolutionary 'red bureaucracy' would simply replace one form of coercion with another, and agrees with him that this kind of socialism *without* freedom is not his kind of socialism either.

Anarchism remains a broad church, but one that is usually divided into three streams: individual anarchism, anarcho-syndicalism and anarcho-communism or communitarian anarchism (see Woodcock 1963; Ritter 1980; Miller 1984; Marshall 1993; Sheehan 2003). As

Sheehan (2003, 38) points out, anarchism's spectrum is well captured 'in terms of an arc that spans communism and extreme individualism'. Kropotkin himself identified six main anarchist schools: mutualist, individualist, collectivist, communist, Christian and literary; and major writers on anarchism agree to a similar typology but with the inclusion of syndicalism (see Kinna 2005, 17, 20). A much simpler distinction can be drawn between 'individual' anarchism and 'social' anarchism since this encapsulates the key elements of anarchist thought and of the different schools within it. Generally speaking, individual anarchists privilege the individual within the community and favour autonomous solutions to social problems. Social anarchists instead favour communal responses to social problems. While viewing the individual as key, social anarchists believe that individual flourishing can only occur in a communitarian society. But both promote, if in different ways, maximum freedom for individual expression in a community that sponsors harmonious relationships with fellow human beings.

Social 'versus' individual anarchism

The 'mainstream' is considered to be occupied by the social anarchists who have always formed the majority. They theorize forms of social organization that best advance the goals of individual freedom and autonomy. Hence collective action is central to both their means and their ends. Amongst the social anarchists we find the key players – the mutualists, collectivists, communists, communitarians and syndicalists – although it is also important to distinguish them in terms of their political philosophy as well as their broad political strategy. While the social anarchists may differ in emphasis – highlighting anarchism's mutualism, collectivism, federalism or its scientific character – they nonetheless agreed to some key principles. They all privileged individual autonomy and freedom; mounted an overarching critique of the state – each seeing it as a corroder of liberty and a destroyer of autonomy; and articulated a non-hierarchical and non-centralized social vision best positioned to promote individual flourishing.

The central communitarian anarchists – Proudhon, Bakunin, Kropotkin – differ quite markedly however from the radical individualist stream of a Max Stirner, and of course the anarcho-capitalist derivatives of classical liberalism. In *The Ego and its Own,* Max Stirner famously comments that 'nothing is more to me than myself', setting in place the fundamentals of his egoistic anarchism (Stirner 1995). Signalling the difficulties of neat categorization, however, Proudhon also mounted a strong individualist case. In *What is Property*, Proudhon argues that the social ownership of property under communism can also endanger individual freedom and choice. This mimics aspects of Stirner's critique of authoritarian communism:

> Communism, by the abolition of all personal property, only presses me back still more into dependence on another, to wit, on the generality or collectivity ... [This is] a condition hindering my free movement, a sovereign power over me. Communism rightly revolts against the pressure that I experience from individual proprietors; but still more horrible is the might that I puts in the hands of the collectivity (Stirner 1995, 257).

If the social anarchists identify in community the mechanisms of individual flourishing, individual anarchists fear in this same community their very loss. Proudhon shares some of Stirner's reservations that community can act as its own form of tyranny. Proudhon was clearly against private property and the capitalist state that supported it, and in his federalist model sought to empower independent, autonomous and decentralized communities. But Stirner's critique of authoritarian communism cannot be read as an endorsement of capitalism either. His 'egoistical' individualism did, after all, reject capitalism, the market and the institution of private property (see Miller 1984, 22–5; Marshall 1993, 220–33).

Stirner has been relatively overlooked in the anarchist pantheon – until quite recently. His egoist focus made it hard to discern a practical politics for transforming the social conditions of oppression and exploitation. In order to protect the individual's integrity, Stirner's hyper-individualism required a 'disassociation' between individuals. He resisted any calls to define the individual's essence, or to socially corral them into communal activities, for fear of trampling their 'uniqueness'. Communists and revolutionaries of the time, including social anarchists, were exasperated at his proposal for a seemingly apolitical 'union of egoists' as a revolutionary beacon. But Stirner was careful to distinguish revolution from insurrection:

> Revolution and insurrection must not be looked upon as synonymous. The former consists in an overturning of conditions, of the established condition of *status*, the state or society, and is accord-

ingly a *political* or *social* act; the latter ... is not an armed rising but a rising of individuals, a getting up without regard to the *arrangements* that spring from it. The Revolution aimed at new arrangements; insurrection leads us no longer to *let* ourselves be arranged, but to arrange ourselves, and sets no glittering hopes on 'institutions'. It is ... a working forth of me out of the established (Stirner 1995, 279–80).

For Stirner, individual insurrection, so understood, is the ultimate political act since its destabilizes 'the established' from the 'inside' rather than just the 'outside'. At a time when most fellow revolutionaries championed modernism's humanism, Stirner argued instead that it drove the individual's enslavement. He saw as the ultimate act of domination and power the inscribing of individual identity. By internalizing the mechanisms of repression, the individual, rather than an external state, exercises domination: 'a master is a thing made by the servant. If submissiveness ceased, it would be all over with [the state's] lordship' (1995, 175). Since individuals 'cooperate' in their own oppression, their ultimate insurrectionary act is to reclaim their desires. Such claims prefigured Foucauldian and Situationist analyses, as discussed in a subsequent section.

Despite Stirner's rejection of capitalism, his 'breed' of hyper individualism was seen as fuelling the individualist anarcho-capitalist position, a position that took firm root in the United States. Social anarchists are particularly scathing of the anarcho-capitalism of the Jeffersonian democrats who developed and extended classical liberalism's idea of the sovereign individual 'until it became incompatible with the idea of the state' (Miller 1984, 30). The US's fight for independence against external hierarchs was underpinned by the desire to create an autonomous society where private property and the ethos of individualism prevailed. Like most other ideologies, anarchism manifested differently in different countries, responding to diverse political cultures and political histories. While individual anarchism, or a US brand of 'libertarianism', took root in the United States, it was an altogether different experience in Europe. European anarchists usually attached a descriptive qualifier to their political brand. They were not simply libertarians, but usually libertarian socialists or some such, keen to distinguish themselves from their US counterparts who readily mixed capitalism with their libertarianism. As Chomsky (in Chomsky & Vodovnik 2004) points out:

The individualist anarchism [of] ... Stirner and others, is one of the roots of – among other things – the so-called 'libertarian' movement in the US. This means dedication to free market capitalism, and has no connection with the rest of the international anarchist movement ... As far as I can see, the workers' movements [in the US] which didn't call themselves anarchist, were closer to the main strain of European anarchism than many of the people in the US who called themselves anarchists ... If we go back to the labor activism from the early days of the industrial revolution ... it's got a real anarchist strain to it. The never heard of European anarchism, never heard of Marx, or anything like that. It was spontaneous.

US libertarians nonetheless shared with European anarchists the privileging of individual autonomy and freedom. Like classic liberals, anarcho-capitalists worry that the autonomy of the individual may be compromised by the dominance of the community and/or the state. However, unlike social or communitarian anarchists, contemporary anarcho-capitalists are the market's loudest champions, arguing for untrammelled market freedom and unlimited private property rights – if attained through (legal) individual endeavour. While communitarian anarchists also uphold the primacy of the individual, they see the integrity and freedom of the individual springing from his/her social relations. The claim is that it is only in the free and mutualist consociations forged in community and in collective interactions that the individual can flourish. And capitalist social relations and the market communities they generate are not conducive to such flourishing. Thus, 'social problems cannot be solved on an individual basis or by the invisible hand of the market'; it is instead 'necessary to combine with others and work together' (Marshall 1993, 10–11). But in refusing to subscribe to consociation and mutualism, commentators such as Marshall hesitate in embracing anarcho-capitalists as 'true' anarchists. This is in part because of the communitarian anarchists' refusal to see human nature as driven primarily by a self-seeking egoism, preferring to temper this egoism with the powerful correlates of sociality and cooperation. This helps explain their distaste for Stirner's 'ego and its own'.

Social or communitarian anarchism is more than a theory of a stateless society, even if it is resolute in its insistence that the state be dismantled. The central argument that the centralized power of the external state must be returned to, and internalized in, the individual-in-community has lost none of its force. While generally rejecting the base and superstructural relationship of the state as espoused by Marxism, communitarians do at least agree with Marx that the capitalist state is an immensely oppressive force. For anarchists, this is the very state that prevents the development of freedom, cooperation and responsibility in human beings. But any group, including Marx's proletarian 'freedom-fighters', that assumes even the temporary reins of statist power will be inevitably corrupted. Anarchists have long argued that it is hierarchy and authoritarianism, whether wielded by the state or a revolutionary vanguard, that corrupts the human spirit and derails freedom and subjectivity. Marx's view of the state was for anarchists far too narrow. Yet the 'early' Marx and the 'early' anarchists shared much.

To the degree that both identified private property as the root of social corruption Marx's early socialism and Bakunin's communitarian anarchism largely concur. Bakunin too was seduced by the Left Hegelians and shared much with the young Marx. It is when each outlined their strategies for the transformation of society that serious rifts began to develop. As the historical record testifies, there was bitter conflict between Bakunin and Marx over control of the First International. Bakunin's walkout prompted the ensuing divisions between communist anarchists and state socialists. While the goals may have been similar – that is, emancipation and empowerment for the oppressed masses – the strategies differ significantly. For Bakunin, there could be only one logical conclusion to Marx's scientific socialism – the supplanting of one form of oppression by another. While Bakunin adopted, and adapted, Marx's economic materialism he developed Marx's class analysis considerably, locating revolutionary potential beyond the working class to the peasants and even the 'dreaded' lumpenproletariat. By promoting the spontaneous uprising of all comers in their clamouring for freedom and autonomy, Bakunin left little room for Marx's scientific socialism, and for the socialist state's paternalistic role (see Malatesta 1974). By the 1860s the divisions between the Marxists and the anarchists were clearly etched, and the anarchists were busy forging their own emancipatory project. The visionary breadth of this project is well exemplified by Kropotkin:

Anarchism is a world-concept based upon a mechanical explanation of all phenomena, embracing the whole of nature – that is, including in it the life of human societies and their economic, political, and moral problems. Its method of investigation is that of the exact natural sciences. ... Its aim is to construct a synthetic philosophy comprehending in one generalisation all the phenomena of nature – and therefore also the life of societies (Kropotkin 1975, 61–2)

Social anarchism proposes a more positive conception of human nature, even as it acknowledges the limits of egoism. It does not posit a conflict-free future but it does submit that an anarchist society would be more able to mitigate the tension between egoism and altruism. It envisions a future that successfully arbitrates individual freedom and community responsibility – or self-interest and cooperation – drawing its insights from anthropology and the evolutionary and ecological sciences. While not denying the impulse of competition in driving evolution, Kropotkin highlighted the equal contribution of its corollary – cooperation. Kropotkin's scientific analysis led him to emphasize the empowering potential of mutual aid and helped lay the scientific foundation on which this influential strand of classical anarchist theory rests. This scientific character helped justify its political analysis and the cultural renewal strategies it promoted. Even so, the theory of competition gained more currency, assisted somewhat by the sophisticated justification it provided for the social Darwinism of a winner-takes-all *laissez-faire* capitalism. As a prominent scientist and social critic points out:

> What can we make of Kropotkin's argument today, and that of the entire Russian school represented by him? Were they just victims of cultural hope and intellectual conservatism? I don't think so. In fact, I would hold that Kropotkin's basic argument is correct. Struggle does occur in many modes, and some lead to cooperation among members of a species as the best pathway to advantage for individuals. If Kropotkin over-emphasised mutual aid, most Darwinians in Western Europe had exaggerated competition just as strongly. If Kropotkin drew inappropriate hope for social reform from his concept of nature, other Darwinians had erred just as firmly (and for motives that most of us would now decry) in justifying imperial conquest, racism, and oppression of industrial workers as the harsh outcome of natural selection in the competitive model. (Gould 1991, 338)

Compared to the Hobbesian view that only an authoritarian state can corral a brutish human nature, the anarchist reading of human potential is far more optimistic. This optimism, at least in social anarchists such as Kropotkin, Bakunin and Bookchin, relies on insights drawn from evolutionary science. Bookchin's social ecology, draws substantially, if critically, from Kropotkin and the communitarian anarchists and offers a similar vision of an enlightened future. As discussed in

Chapter 6, for Bookchin the 'good' society becomes the ecological society, thus the neologism: eco-anarchism.

The tensions and debates between anarchism and Marxism, and between the social and individual anarchists, simmered over a number of tumultuous decades. Anarcho-communism and anarcho-syndicalism were sustained by the pressures communism continued to exert on its capitalist nemesis, as well as by the successes of the Spanish anarcho-syndicalist experiments of the mid-1930s. But when the Stalinist excesses and communism's authoritarian tendencies were finally exposed, the communist left was both humbled and humiliated. An emergent group of critical theorists then led the left through a painful debriefing process, paving the way for the significant reconceptualization of core principles – a reconceptualization with a decidedly anarchist flavour.

Anarchism, the New Left and beyond

From the outset, anarchism railed against the authoritarian excesses of Marxism and demanded a broader understanding of the sources and operation of power. It drew the 'best' from competing ideologies, including liberalism, to assemble its utopian sweep. When the Soviet communist project began crumbling, well before 1989, the New Left began reassessing leftist orthodoxy (Fromm 1942; Marcuse 1964; Habermas 1970; Adorno & Horkheimer 1972; Adorno 1973; Horkheimer 1974; Fromm 1976). In the process they drew considerably from the insights of anarchism. Herbert Marcuse's *One Dimensional Man* (1964) and Jurgen Habermas's *Towards a Rational Society* (1970) proved particularly compelling in their critique of industrial society. In demonstrating the ubiquitous penetration of capitalism in all areas of lived experience, they contended that its dominance could only be challenged by a transformed, post-Marxist approach to social change. The postmodernists and poststructuralists that followed them went considerably further in condemning the whole modernist project itself. Modernity's political quest was nothing less than the transformation of human society, typified by such revolutionary slogans as 'liberty, fraternity, equality'. But the universal ideals and universal conceptions of the 'good' that drove modernism were, according to postmodernists, underpinned by an instrumentalist and rationalist ethos. Instead of freedom, the enlightenment instead drove the technical mastery of humans and nature. Now that 'truth' had been discovered and the 'common good' defined, all that remained was the translation of these principles into rational social models. The modernism that resulted was positivistic, technocratic and rationalistic.

Viewed in this way, postmodernism was a rebellion against the anthropocentric, androcentric and ethnocentric sensibilities of modernism. These insights collected in the new social movements and particularly the green movement, as well as stimulating new theoretical developments within anarchist thought itself. Increasingly, modernism's universal and essentialist claims were challenged, as was the instrumental rationality that drove them. As Harvey (1989, 354) notes, the New Left saw itself 'as a cultural as well as political-economic force, and helped force the turn to aesthetics that postmodernism has been all about'. In expanding the domination discourse beyond the Marxist one, the New Left named instrumental rationality as one of capitalism's key tools of domination. Essentialism and instrumentalism began to be very directly linked to the domination of both people and nature.

Much anarchism felt vindicated by such analyses. Others, such as Murray Bookchin, saw in the theorization of postmodernism and poststructuralism, empty obfuscatory rhetoric designed to mask rather than illuminate domination. Nonetheless, the resonance between anarchist and, for example, Foucauldian analyses illustrates well the contributions of postmodernism and poststructuralism to the broader discourse of domination. Even among anarchists distrustful of these new currents, there is a shared identification of a 'logic' and practice of domination that wreaks considerable social and environmental havoc. After Foucault, there is also a shared recognition that power is not simply an expression of sovereignty but manifests in a variety of disciplinary practices. Foucault views the institutional state as heavily implicated in the fixing of domination but also locates a myriad of other sites in which power is exercised (Foucault 1969, 1976). Foucault's identification of individual repression in a multitude of political, social, sexual, cultural and emotional practices, and his claim that repression is both the means and ends of power, caught the sharp attention of anarchists.

The convergences between anarchist views and those of other poststructuralists have been highlighted more recently in the work of 'postanarchists', as discussed in a subsequent section. In rejecting a deterministic and class-reductionist view of power, and in arguing that there exists a myriad of micro-technologies of power with 'totalizing' effects, Foucault locates the sources of this power as both within and outside the state. To the degree that it is resourced with very potent 'technologies of power', the state is well situated to effect 'relations of force' that well serve its objectives of social control. Yet the logic of domination is not confined to the institutional state alone. As many critical theorists argue, and many anarchists would concur, the centralized power of the state enables it to strengthen the underlying mechanisms of repression: hierarchy and oppression. The modern state has an unprecedented capacity for power and control, especially through its facility for surveillance and regulation. Its 'relations of power' are now ubiquitous and take many forms. As Foucault claims:

> One impoverishes the question of power if one poses it solely in terms of legislation and constitutions, in terms solely of the state and the state apparatus. Power is quite different from and more complicated, dense and pervasive than a set of laws and a state apparatus. It's impossible to get the development of productive forces characteristic of capitalism if you don't at the same time have apparatuses of power (Foucault 1980, 158).

Despite a resonance with anarchism, Foucault is no anarchist and would resist such a label. His anarchist proclivities reside primarily in his suspicion of statist solutions and his endorsement of local, decentralized and particularized social struggles. He believed it crucial that such struggles remain outside the ambit of the state if they were to avoid the institutionalized absorption that would otherwise occur. These decentralized, site-specific and often single-issue struggles against 'particularized power' – prisoners, women, gays, hospital patients and so on – need seek no unity or solidarity with other dissidents. In short, they should remain decidedly *autonomous*. Nonetheless, Foucault's tendency to view the 'will to power' and associated repressive practices as endemic to human nature and as evidence of the 'immanence' of power, represents the point at which many anarchists demurred, agreeing with Marxists that repression is also very much structurally determined. Foucault's eschewal of utopian or visionary politics also exasperated anarchists (Marshall 1993, 586). In his world of moral relativities, Foucault's refusal of a transformational politics and social change alternatives still induces considerable anarchist scepticism. In a televised debate with Chomsky in 1971, Foucault refused to be drawn on preferred social models, arguing instead that 'the task of the revolutionary is to conquer power, not to try and bring about justice which is merely an abstraction' (cited in Marshall 1993, 586).

Overall, the influence of the new epistemological currents of critical theory, postmodernism and poststructuralism on the shape and politics of the new social movements accommodated an increasingly confident anarchism. Many anarchists were particularly heartened to see the critique of domination extend beyond the social relations of capitalism to include the politics of race, gender, culture and nature. It was equally heartened by the increasing adoption of direct action as a key political strategy. This is not to say that the new social movements eschewed more traditional forms of protest altogether – this was after all the era of the non-government organization, many of which were hierarchically structured, tightly organized and highly professionalized. Nonetheless, through the adoption of analyses and strategies that resonated anarchist impulses, the new social movements were successful in placing issues of gender, environment and race on the mainstream political agenda – in the developed north at least. The politics of inclusion hence empowered the 'other voices' that Marxism and liberalism did not embrace, and that anarchism had long championed. No longer was the proletariat, or even the labour movement more generally, to dominate the politics of dissent. In short, anarchism – and postanarchism after it – drew considerable strength from Foucault's proclamation that:

> ... power in Western capitalism was denounced by Marxists as class domination; but the mechanics of power in themselves were never analysed. This task could only begin after 1968, that is to say on the basis of daily struggles at the grass roots level, among those whose fight was located in the fine meshes of the web of power (Foucault 1980, 116).

New anarchism: post-leftism

Vibrant, reflexive ideologies continually remake themselves in the light of contemporary socio-political developments and the new challenges they raise. Anarchism is no different. The significant developments in anarchism over the past few decades signal a preparedness to reconsider its ideas and praxis in the light of new political challenges. The new anarchism is not a wholesale rejection of old or classical anarchism. But it draws from and reconfigures itself in new ways, even as old ideological battles, fought under the rubric of new forms, are still clearly visible. In articulating new positions, it blends old with new or rejects one in favour of the other – all, albeit, with varying degrees of coherence and success. One tension that has been visibly resurrected is that between individual and social anarchism. Others include different views on the ethos and value of technology and the utilization of post-structuralist theories in the informing of new anarchism. Many of these tensions emerge within a discourse of 'post-leftism', a broad response to the post-1989 political climate in which contemporary dissent is situated.

Post-leftism is not so much an ideological position as a political principle that underpins much new anarchism. Under its broad umbrella we can include some poststructuralist anarchisms, anarcho-primitivism (although not all post-leftists are primitivists) and Bey's ontological anarchism, each discussed in turn below. Bob Black is often identified as a key defender of post-leftism and his *Anarchy After Leftism* (1997) is considered a seminal work in the area. His condemnation of work (1986) as a site of discipline and control is probably better known, however. As a broad idea or tendency, post-leftism highlights an important development in contemporary anarchist theory and practice. Post-leftism rejects both the politics of the broad left and ideology itself. The claim is that the brutality of the communist experiment exposed an ideology driven by a relentless and repressive architecture. Anarchists had always warned against the dangers of such authoritarianism, but social anarchists, by and large, still upheld the principle of collectivism, ignoring, according to Black, its inherent capacity to nurture authoritarianism. Post-leftism thus denounces both ideological authoritiarianism and the collective organizational culture in which it breeds. Class, collective and revolutionary organizations were often transformed into the new tools of domination, eventually overtaking the repressive organizations they sought to replace. Black (1997, 61) finds in organizations themselves that 'the means tend to displace the ends' and 'the division of labour' that underpins these organizations 'engenders inequality of power'. Instead of traditional revolution, post-leftists promote insurrection. Hakim Bey, another contemporary anarchist, proclaims that it is important not only to give up 'waiting for the revolution' but also to stop 'wanting it' (Bey 2003, 99). The insurrectionist act is spontaneous and autonomous and not beholden to the organizational strictures of revolution and vanguardism.

Once again, Black labels post-left anarchism 'type-3 anarchism' – a decidedly non-ideological form of anarchism that he claims is neither individualist nor social in its entirety. Post-leftists are also labelled 'egotist-anarchism', reflecting their roots in the anarcho-individualist tradition. Rather than a 'school' or movement – organizations they would oppose in any case – post-leftism repositions individuals and autonomous insurrectionist acts at the centre of their transformational politics. Still, they would bristle at accusations that they are 'simply' individualist anarchists. They recognize, after all, that individual autonomy can only be properly realized in a free community. Another post-leftist, Jason McQuinn (2003), contends that critics 'simply don't understand the huge divide between a self-organising movement seeking to abolish every form of social alienation and a merely political movement seeking to reorganise production in a more egalitarian form'. Post-leftists object to a 'false consciousness in which people no longer see themselves directly as subjects in their relation to the world' (2003). Hence they claim to be 'neither left, nor right, but auto-nomous', a movement 'that can stand on its own and bow to no other movements' (2003).

Bey helps illuminate what a post-left anarchism might mean. He identifies the lesser stars among the pantheon of anarchist theorists from which they draw. For post-left or type-3 anarchists 'Stirner out-weighs Marx'; 'Fourier was amazing' and into their fold they have welcomed 'the Ranters, Antinomians, and Family of Love, as well as radical forms of Buddhism, Taoism and sufism' (Bey 1991). McQuinn contends that contemporary ideological anarchism is obsessed with repeating the grandeur of past struggles such as Spain 1936 and Paris 1968. But to remain relevant and, more importantly, to continue serving as an inspirational force for radical change, anarchism needs to 'address the lived realities of the twenty-first century ... leaving the outmoded politics and organizational fetishism of leftism behind' (McQuinn 2003). While postanarchists might construct their radical positions differently, there is considerable empathy with the post-left position. Newman claims, for example, that there is:

> ... a certain litany of oppressions which most radical theories are obliged to pay homage to. Why is it when someone is asked to talk about radical politics today one inevitably refers to this same tired, old list of struggles and identities? Why are we so unimaginative politically that we cannot think outside of this 'shopping list' of oppressions? (Newman 2001, 171)

We turn to recent theoretical developments that, to various degrees, construct post-left anarchist positions – positions that resonate elements of post-ideological anarchism.

Anarchism and poststructuralism: recent developments

Poststructural anarchism is not a formally developed 'school' of anarchism. Rather it proposes that by drawing insights from a broader range of poststructuralist theorists, anarchism can be made far more relevant and appealing to a new generation of activists. Poststructuralist anarchism utilizes the insights of such poststructuralist luminaries as Lyotard, Deleuze, Lacan and, as we have already noted, Foucault. In his *The Political Philosophy of Poststructuralist Anarchism* (1994) Todd May denies that there is, or should be, a single agency or mode of social transformation. His objective is to remove anarchism from the universalist limitations that still contain it. While capitalism and the state are important oppositional targets, he argues that anarchists need to go beyond such singular sites of power if effective social change is to be realized. Lewis Call's *Postmodern Anarchism* (2003) draws primarily from Nietzsche, Baudrillard and some cyberpunk literature to construct a society underpinned by radical 'gift-giving' – or the giving of gifts without traditional expectations of return (Glavin 2004). As we note in Chapter 3, Call's analysis uncovers information sharing capacities in the new technology that can be counter-appropriated by anarchists for radical goals. But, like his fellow postanarchists, Call too advocates the revolutionary potential of a politics of desire. Saul Newman's postanarchism also draws from a range of poststructuralist insights to make his case for a renewed anarchism. In his *From Bakunin to Lacan: Antiauthoritarianism and the Dislocation of Power (2001)*, and subsequent writings, Newman sets out to disencumber anarchism of its essentialist and universalist tendencies by remaking it along poststructuralist lines. But the 'post' inscription has two general meanings. The first refers to a shorthand combination of *post*-structuralism and *anarchism*; and the second uses it to denote an *anarchism* that goes beyond *(post)* the ideological limitations that have restrained it.

More generally, the term originates from Hakim Bey's conception of 'post-anarchism anarchy' where he argued that anarchism's doctrinal qualities were, paradoxically, driving away those most committed to its principles of autonomy and anti-authoritarianism. Operating canonically as an ideologically constrained 'sect', and busy fending off all kinds of theoretical challenges, this ideologically corralled anarchism alienated those wanting to make it more relevant to the transformational politics of the new century. Key to this transformational potential is the capacity to admit more players into the mix of radical dissent. Poststructuralism's 'politics of difference' is just the conduit to expand such a mix. Bey's post-anarchy anarchism thus implores anarchists to move beyond comfortable doctrinal purities by embracing the insights of new theoretical developments in the deconstruction of power. This will in turn permit the embrace of a much broader range of struggles and rebels into the anarchist fold. Classical anarchists, after all, chastised Marx for limiting revolutionary struggle to the working class. Contemporary anarchism should avoid likeminded exclusiveness by inviting modern day 'lumpens' into their fold.

Both May and Newman seek to bring the insights of some poststructuralists directly to anarchist theory. When asked how he would reconcile anarchism with a poststructuralism that theorizes multiple powers and has no seeming overarching transformational goal, May responds that while poststructuralism helps fill some of anarchism's analytical gaps, so too does anarchism provide poststucturalism with the means to transcend its own limitations, what he calls a 'larger framework in which to situate its specific analyses'. Hence while anarchism borrows from poststructuralism insights on the operation of power, anarchism provides it with a more viable political form:

> We need to understand power as it operates not only at the level of the state and capitalism, but in the practices through which we conduct our lives ...While [anarchists] have a two-part distinction: power (bad) vs nature (good), I have a four part one: power as creative/power as repressive and good/bad. I do not take creative power as necessarily good, nor repressive power as necessarily bad. It all depends on what is being created or repressed (May in DeWitt 2000).

May is concerned to show how representation snatches and dilutes the individual's power. A transformational politics – which he claims post-anarchism to be – challenges the politics of representation and understands that power is multiplicitous. Like Newman, he identifies an essentialist tendency as classical anarchism's central weakness, although how he reconciles a non-essentialism with an anarchist practice that is 'universal in scope' becomes more problematic (see Glavin 2004, 3).

Sharing similarities with both Bey's 'post-anarchism' anarchy and May's poststructural anarchism, Newman fleshes out his postanarchism considerably. He traces anarchism's modern (or post-modern) transformation from an ideologically rigid classical anarchism to an open and diverse configuration underpinned by the insights of poststructuralism. He claims that:

> ... by using the poststructuralist critique one can theorize the possibility of political resistance without essentialist guarantees ... [and] by incorporating the moral principles of anarchism with the post-

structuralist critique of essentialism, it may be possible to arrive at an ethically workable, politically valid, and genuinely democratic notion of resistance to domination (Newman 2001, 158).

In short, this births an anti-authoritarian postanarchism suitable to its times. Newman (2003) argues that what distinguishes postanarchism from classical anarchism is the former's 'non-essentialist politics', with classical anarchism limited 'by its epistemological anchoring in the essentialist and rationalist discourses of Enlightenment humanism'. Postanarchism's release from such essentialist limitations frees it to engage more effectively with 21st century politics. His claim is that if contemporary anarchism is to prosper, it:

> ... must follow its insight about the autonomy of the political dimension to its logical implications – and see the political as a constitutively open field of indetermination, antagonism and contingency, without the guarantees of dialectical reconciliation and social harmony (2003, 4).

Newman finds that the anarchist impulse embedded in the anti-globalization movement has heeded a Beyian call to 'de-doctrinalize'. Indeed, it is the 'openness to plurality of different identities and struggles, that makes the anti-globalization movement an *anarchist* movement' – but a modern anarchist movement that incorporates some of the non-essentialist insights of poststructuralism (Newman 2003, 1). A non-essentialist anarchism necessarily rejects some of the more essentialist arguments of some of the classical anarchists. Other anarchists, particularly Stirner, are rescued from relative obscurity to occupy a more central place in his postanarchism. It is Stirner's critique of modernism's essentialism and his prefiguring of poststructuralist analysis and the 'politics of difference' that most impresses Newman. Stirner's greatest insight was to identify in the humanist abstraction of 'Man' a replacement for the abstraction of 'God'. Through creating a human essence when in fact there was none, both God and Man denied the individual their uniqueness and integrity. Accordingly, when the state demands 'that I be a man ... it imposes being a man upon me as a *duty*'; to not assume the mantle of such 'manhood' is to invite punishment and estrangement, so that the 'fear of man has taken the place of the old fear of God' (Stirner 1995, 161, 165). Essence or essentialism hence becomes 'the discourse through which political power operates' (Newman 2002, 228). It is only from a non-essentialism, or emptiness, that an individual can construct his/her identity. To regain their power, individuals need to shake themselves loose from externally imposed subjectivities and begin afresh. Newman thus contends that Stirner 'was one of the first to consider the question of self-repression' and the state's reliance 'on our willingness to let it dominate us' (Newman 2002, 229). For Stirner, the control of desire becomes one of the state's ulti-mate weapons, and conversely human liberation lies is the reclamation of desire.

For individualists the retrieval of individuality – one's ego – becomes the ultimate revolutionary act. But this 'insurrection of the self' incites a very different kind of revolution. 'Ordinary' revolutions simply trans-plant one form of authority for another: the concept of authority itself is never properly challenged. As Newman observes, Stirner discerned that revolution was simply 'the imposition, in the name of universal-ity, of one identity, one particularity on others' (2002, 232). Stirner's true contribution thus lay in his articulation of 'a series of strategies which embrace plurality over essence, ethical questioning over moral and rational certainty, and contingency of identity rather than an oppressive stability' (2002, 238). Certainly, these qualities have pre-figured aspects of poststructuralist theorization, as well as inspiring the shape of much contemporary anarchism.

Contemporary anarchism can thus be viewed as both '*post*'-anarchist and post-'*anarchist*' in that 'it is not a complete rejection of classical anarchism but rather a step beyond the limits defined for it by Enlightenment thought' (Adams 2004). In short, through using the insights of poststructuralism, postanarchism seeks to rescue anarchist theory – and in particular its core principle of autonomy – from the political and theoretical scrapheap that a continued attachment to essentialism drives it. But this is no obtuse development since anar-chism can in any case be seen as *in nuce* poststructuralist: 'in its pol-itical orientation [poststructuralism is] fundamentally anarchist – particularly its deconstructive project of unmasking and destabilizing the authority of institutions, and contesting practices of power that are dominating and exclusionary' (Newman 2003, 5). To this degree, postanarchism is a hybrid theory that brings together the 'best' of a number of theories to create an emancipatory radical politics capable of resisting dominant powers set on crushing autonomy and diversity (see Morland 2004).

Postanarchism goes a considerable way towards capturing some of the elements that have filtered into our post-ideological anarchism, but on its own it does not go far enough. This is because there is also a broader constellation of influences from which a post-ideological anarchism draws. This includes the contributions of radical ecology, Situationism, postcolonialism, autonomism, postfeminism and Zapatismo – as well as those from classical anarchism and Marxism. Our next new anarchist position, anarcho-primitivism, is considerably different in flavour, even as it claims a strong post-leftist sympathy. It is not enamoured of postmodernist or poststructuralist ideas, and takes anti-authoritarianism to giddy new heights. Its following might be limited, but its stark proclamations resonate in the corridors of radical dissent.

Anarcho-primitivism

'John Zerzan doesn't have a car, a credit card or a computer. He lives a quiet life in a cabin in Oregon and has sold his own blood plasma to make ends meet'. So opens an interview by *The Guardian's* Duncan Campbell with Zerzan in 2001. Zerzan – the infamous, anti-civilization American anarchist who has claimed Unabomber Ted Kaczynski among his influences – is one of contemporary anarchism's most controversial figures. Zerzan's anti-authoritarian anarchism is guided by a passionate loathing for civilization or 'symbolic culture'. This loathing extends to an authoritarian state that alienates humanity in the name of civilization. But he contends that it is not enough to resist the modern state. It is civilization itself that needs to be opposed. The state simply disguises itself in the sophisticated technological, consumerist and cultural garb of civilization. Zerzan's anarcho-primitivism would prefer to dismantle civilization, but short of that, seeks to live an autonomous, 'rewilded' life as much outside civilization as possible. He finds that the roots of domination and hierarchy lie in the division of labour and specialization of tasks generated by the rise of agriculture, surplus production and, finally, industrial society. These developments displaced the egalitarian and non-hierarchical hunter-gatherer societies who lived harmoniously with nature. The division of labour demanded by agricultural production established the ingredients of a social hierarchy that eventually unleashed both social and environmental ruin. Anarcho-primitivism promotes a unique solution to the corrosiveness of civilization: a process of 'rewilding' – of becoming 'uncivilized'.

While not all primitivists pursue rewilding or a 'return to nature', the more radical amongst them make stark proclamations. The anarcho-primitivist journal *Green Anarchist* aspires to nothing less than 'the destruction of Civilization'. For such radicals, civilization remains culprit in social and environmental devastation, representing the 'logic, institutions, and physical apparatus of domestication, control and domination' (Green Anarchy Collective 2004, 36). Civilization privileges symbolic thought and mediates all forms of communication and interaction. The privileging of the symbolic is 'a movement from direct experience into mediated experience in the form of language, art, number, time, etc.' (2004, 36). The damage symbolic culture does is in the filtering of 'our entire perception through formal and informal symbols' so that the world begins to be experienced purely through 'the lens of representation' (2004, 36). Even Chomsky, wearing his other hat of renowned linguistics professor, is condemned by Zerzan for submitting that language development is an inherent, distinctive and marvellous human trait. The primitivist sympathizer Bob Black concurs, condemning Chomsky's 'scholastic obtuseness' and 'Cartesian linguistic theory'. Chomsky's greatest sin seems to be his refusal to see language, particularly written language, as the prefigurement of modern technology and hence 'an instrument of domination' (Black 2004, 6).

Zerzan marshals a wealth of anthropological data to support his case. He intends to dispel the myth of the brutal savage and demonstrate that 'life before domestication/agriculture was in fact largely one of leisure, intimacy with nature, sensual wisdom, sexual equality, and health ... a state that did not know alienation and domination' (Zerzan 1994). The great lie was to present modern civilization as the acme of progress, whereas true progress lay in primitivism or 'uncivility'. But anarcho-primitivism also draws on a broader set of influences. John Moore (n.d.) suggests that anarcho-primitivism derives its inspiration from a range of sources, including animal liberation, feminism, Situationism, deep ecology and Ludditism, but also goes beyond them to propose an immensely radical anti-civilizational alternative. An unyielding anti-authoritarianism and a pursuit of the principles and practices of egalitarianism and mutual aid, gives primitivism its anarchism. But as self-described *green* anarchists they also challenge the domination of non-human as well as human nature, drawing from deep ecology its anti-instrumentalist biocentric ethic. Their neo-Ludditism provides them with their anti-technological impetus and a direct action ethos. They hone from Situationism a vigilance against commodity culture's resolve to corrupt dreams and mediate desire.

The radical anti-authoritarianism at the heart of primitivism seeks to recover individual integrity and autonomy against a civilization-induced alienation that derails subjectivity, creativity and individuation. This objective reflects primitivism's strong anarcho-individualism.

But it is the aversion to technology (and civilization) that sits at the centre of anarcho-primitivism. For Zerzan, while Marx may have correctly identified alienation as a tool of repression, he overlooked the central role of technology in driving this alienation. Technology's greatest sin is to sever individuals from their fundamental nature as creative, autonomous, playful and nature-attuned human beings. He argues that:

> We have taken a monstrously wrong turn with symbolic culture and division of labor, from a place of enchantment, understanding and wholeness to the absence we find at the hear of the doctrine of progress. Empty and emptying, the logic of domestication with its demand to control everything now shows us the ruin of the civilisation that ruins the rest. Assuming the inferiority of nature enables the domination of cultural systems that soon will make the very earth uninhabitable (Zerzan 1994, 15)

The turn to agriculture, industrialism and then capitalism are all driven by a technology that degrades both people and nature. While anarcho-primitivists are usually anti-capitalist – they argue that capitalism and globalization are after all driven by rapacious technologies – capitalism simply represents another, albeit very sophisticated, production process that is to be opposed. Regardless of the era, however, it is the reproduction process itself that drives domination (Perlman 1969).

Anarcho-primitivism opposes not only all systems and institutions but all power relations. Since technological systems are the most insidious in their capacity to dominate and destroy they must be opposed in their entirety (Moore n.d.). They see technology as highly ideological, reflecting the values and goals of those who produce and wield it. It not only drives civilization but constructs it in the image of its masters. Technology thus becomes:

> ... the sum of mediations between us and the natural world and the sum of those separations mediating us from each other ... [and the source of] all the drudgery and toxicity required to produce and reproduce the state of hyper alienation we languish in. It is the texture ... of hierarchy and domination (Moore n.d.).

A civilization tooled by technology translates to a 'global domestication machine' fuelled by a modernity that is 'inherently globalizing, massifying, standardizing' (Zerzan 2004, 16). But technology is not to be confused with tool-making: tools enable straightforward tasks but technology seeks to dominate and control.

While striving for egalitarian communities, primitivists are more concerned with the corrosive impact of civilization and technology on individual autonomy than with broader social consequences. Anarcho-primitivism's strong individualist ethos sees it pursuing the Situationist goal of releasing desire from the stranglehold of civilization. While many primitivists are involved in mass anti-capitalist actions, they are at the same time careful to not be swept up in collective organizations and/ or class actions. Their claim is that, even with the best of intentions, the collective 'organizational model suppresses individual needs and desires for the "good of the collective" as it attempts to standardize both resistance and vision' (Green Anarchy Collective 2004). In short, individuals need to be ever vigilant of all kinds of attempts to usurp their autonomy. While supportive of affinity groups, even these are 'best left organic and temporal', never 'fixed and rigid' (2004). This refusal of any form of power thus sees them suspicious of long-standing forms of collective action. It also renews objections to the Marxist productivist ethic that, while rejecting capitalist production, extolled industrialism as the source of abundance, a view that also resonates deep ecology's objection to a Marxist politics.

In short, anarcho-primitivists take to an extreme the general anarchist objection to mediation and representation. They see in symbolic culture, including language and art, a betrayal of direct unmediated experience and its replacement with separation and alienation. Such a collection of austere views places primitivism at the extreme end of the radical ecology/anarchism spectrum. For some critics this simply provides more ammunition for its dismissal. Even if they agree with its spirit, they have difficulty discerning from it a practical, transformational politics. Many critics also identify in it a decidedly misanthropic undercurrent. They especially abhor its seeming dismissal, in one broad sweep, of the human potential for creativity, innovation and endeavour. They dismiss as fanciful and hopelessly romantic primitivism's seeming championing of a cornucopian return to nature.

Yet primitivism identifies a disillusionment with materialist values that is relatively widespread and that has penetrated the psyche of much oppositional politics. As Smith (2002b, 407) argues, primitivism's 'emergence signifies contemporary disaffection with the ideology of "progress" so central to modernity and capitalism' and identifies a 'culture of contamination' which many radicals similarly identify and decry. For primitivists the only way to avoid this culture of contamination is the establishment of independent non-materialist communities that privilege, above all else, a non-commodified individuation and unmediated relations (including with nature). For many similarly disenchanted fellow radicals, the alternative of 'rewilding' goes too far. But for drawing sharp attention to the perilous path a capitalist civilization drives culture and nature, it wins many friends; for rejecting technology altogether, even for subversive purposes, it deters them. The next view promotes a significantly different approach, but one that is still underpinned by an aversion to the trappings of commodified culture.

Temporary autonomous zones

Hakim Bey is a pseudonym for the writer Peter Lamborn Wilson who has published much material under both names but is best known for his 'ontological anarchism' and the 'temporary autonomous zone' (TAZ). Bey's concept of the TAZ has been enthusiastically embraced by anti-globalization protest and was particularly influential in shaping Reclaim the Streets (RTS) actions. TAZ is an oppositional tactic that pursues a temporary occupation of space, or autonomous zone, for a short escape from the state's institutionalized control. In this free space, individuals can express themselves autonomously, creatively and spontaneously. The TAZ is thus 'like an uprising which does not engage directly with the State, a guerrilla operation which liberates an area (of land, of time, of imagination) and then dissolves itself to re-form elsewhere/elsewhen, *before* the State can crush it' (Bey 2003, 99). TAZ 'uprisings' are both spontaneous and unpredictable, with no ongoing organizational structure directing them. In this way, the TAZ discards the role of a revolutionary vanguard and a coordinating organizational culture, and frees itself from the tactical singularity of revolutionary politics. Ontological anarchists support the idea of uprising but not revolution since the latter implies planning and centralization while the former is non-hierarchic and spontaneous. Transient TAZ uprisings are always preferred over more permanent, organized forms. Indeed, the TAZ 'is a revolution that fails, but only because success would be the ultimate failure, the denial of future TAZs' (Truscello 2003, 13).

Ontological anarchism ventures the quasi-nihilist claim that chaos is the only certainty afforded us. Instead of challenging the long-held myth that 'anarchy is chaos' Bey instead champions it. This invokes the chagrin of many anarchists anxious to deny the perceived link between anarchy and chaos, but for Bey, life is chaos and chaos is life.

Indeed, '[a]ll mess, all riot of colour, all protoplasmic urgency, all *movement* – is chaos. From this point of view, Order appears as death, cessation, crystallization, alien science' (in Bey 1993). The state's great sin is the attempt to create order where there can be none, since any 'form of "order" which we have not imagined and produced directly and spontaneously in sheer "existential freedom" for our own celebratory purposes – is an illusion' (1993). Ontological anarchism thus proposes that individuals shape their own contexts or create their own 'day', even 'in the shadow of the state'. Bey seeks ultimately 'the possibility of creating a chaotic situation, a creative chaos in which there is no centre, but in which there is a multitude of relations between more or less equal powers' (Sugar 1995). It is primarily through this reclamation of desire that the state is usurped and individuation enacted. A totally autonomous moment, experienced within a temporary autonomous zone, becomes the nascent activation of desire. Since only desire creates values, and 'the values of Civilization are based on a denial of desire', only in moments not circumscribed by the state or civilization can true desire, and hence real values, flourish (Bey in Sugar 1995). This is ultimately the TAZ moment.

Described by Bey as a type of 'post-anarchism anarchy', ontological anarchy was particularly influential in rave circles in the 1990s (see Luckman 2001b). Its extensive influences include 'Sassurian semiotics, Hegel, Foureir, The SI, Bataille, Deleuze, Guttari, Lyotard, Thoreau, Bakunin, Nietzche, McLuhan, Virilio, the Surrealists, Baudrillard, Foucault, Kropotkin and Chomskyan Linguistics' (Luckman 2001b, 61). Its neo-Situationism is exemplified both in ideological content and style of delivery. Bey's articulates his polemics in a highly poetic, lyrical and stream of consciousness style that seeks a literal union between art and everyday experience. In doing so he rails against the alienation that drives capitalism and disrupts subjectivity. Human essence is creative, spontaneous and ultimately autonomous. Rather than seizing power through an organized revolution – the route of 'traditional leftism' – he subscribes to a Situationist revolution of the imagination. In articulating his politics of the dance, the rave and the party, Bey draws from Pearl Andrew's conception of the dinner party. For Andrews (1852) the dinner party acts as metaphor for the highest form of human engagement. It is an activity where a group of equal individuals meet for the free, spontaneous and mutually respectful sharing of pleasures and conviviality. Bey too contends that in the conception of the party – whether a rave, RTS party or similar event – can be located the very ingredients of freedom.

Bey advocates a position he calls 'non-hegemonic particularism' – a form of 'radical tolerance' akin to radical pluralism. Unlike capitalism and communism, that both promote a single ideology and culture, a non-hegemonic particularism would instead generate a 'true diversity'. He uses a Zapatista example to illustrate his case:

> We can look at the Zapatistas in Chiapas, Mexico, where there are people who are saying, look we are Mayan Indians, we want to be Mayan Indians and have our own culture, our own personality, our own personhood, we want our own language taught in our own schools, we don't want to become Mexicans or North Americans, we want to be Mayans, but we're not telling you to be a Mayan, we want the freedom to be who we are, we're not telling you that you have to be like us. That's what I call non-hegemonic particularism (in Sugar 1995).

By implication 'non-hegemonic particularism' can only take place in autonomous spaces that permit the articulation of individuals 'desires. These spaces range from RTS parties, Zapatista autonomous zones, social centres (squats appropriated as hubs of oppositional activities), to the autonomous political spaces hosted by World Social Forums. Key to these spaces is their autonomy and their plurality – an autonomous space where 'all worlds are possible' and where each can chase his/her dream. In promoting the notion of autonomous space as central to oppositional politics, Bey champions a politics that is unmediated, localized, non-hegemonic and particularistic. These diverse autonomous politics then meet up to power the movement of movements.

Old tensions revisited

Some social anarchists accept none of the new anarchist 'posturing' – post-leftist, poststructuralist, post-ideological or otherwise – dismissing it all as pretentious bluster. Social ecologist and eco-anarchist Murray Bookchin does just this. In doing so, he resurrects the long-standing tension between individual and social anarchism. In his scathing critique of contemporary anarchism, Bookchin (1996) laments the turn to individualist anarchism – or what he labels (pejoratively) 'lifestyle anarchism'. Bob Black's counter-attack in *Anarchy after Leftism* (1997) is equally scathing. For Bookchin lifestyle anarchism has few valid claims to anarchism's great tradition; it is instead a self-indulgent capitulation to the rampant individualism that underpins the neo-liberal onslaught.

He describes it as an egoistical anarcho-individualism preoccupied with 'polymorphous concepts of resistance' such as '[a]d hoc adventurism, personal bravura, an aversion to theory ... celebrations of theoretical incoherence (pluralism), a basically apolitical and anti-organizational commitment to imagination, desire, and ecstasy, and an inherently self-oriented enchantment of everyday life' (Bookchin 1996). He condemns a considerable cast. These include Hakim Bey and his 'narcissistic anarchism', Zerzan and his 'regressionist' primitivism, 'anti-rationalist' postmodernist anarchists of many persuasions, and the deep ecology-biased radical 'misanthropes' found in Earth First! and beyond. Through the last he resurrects many of the vitriolic debates of the 1980s between the deep and social ecologists, particularly with George Bradford from *Fifth Estate*. He also dismisses anarchist journals and magazines that promote new anarchism – *Fifth Estate, Anarchy: A Journal of Desire Armed* and *Green Anarchist*. In short, he asserts that:

> ... what passes [today] for anarchism in America and increasingly in Europe is little more than an introspective personalism that denigrates responsible social commitment; an encounter group variously renamed a 'collective' or an 'affinity group'; a state of mind that arrogantly derides structure, organisation, and public involvement; and a playground for juvenile antics (Bookchin 1996).

For Bookchin lifestyle anarchism threatens the very integrity of anarchism. Classical anarchism's failure to clearly articulate the relationship of the individual to the collective, has now unleashed a more virulent strain of individual anarchism. As discussed above, classical anarchists have always grappled with this tension, a tension fundamental to anarchism's simultaneous commitment to individual autonomy and collective freedom. Bakunin long contended that only a free society could guarantee the individual's autonomy. The eventual dominance of social forms of anarchism, particularly communist anarchism and anarcho-syndicalism, and the increasing negative association of individual anarchism with anarcho-capitalism, marginalized individualist schools considerably – until comparatively recently. Fundamental to this tension, as now, are the different conceptualizations of autonomy and freedom. Bookchin contends that individualist anarchists privilege the former while social anarchists the latter. For social anarchists such as Bookchin, the needs of the individual are 'dialectically' interwoven with the needs of the collective – each depends on the other. Individualism simply privileges the individual's self-sovereignty, or autonomy, but social freedom enables it.

Bookchin's gripe is with what he considers a wilful distortion of the great concepts of autonomy and freedom that underwrite anarchism. This distortion transforms anarchism into a philosophy that promotes personal autonomy rather than social freedom. Lifestyle anarchism's self-absorbed individual has little time for community, or for the collective decision making on which a free community depends. Social freedom is thus sacrificed at the altar of rampant selfhoods. As a proud municipalist that promotes decentralized and self-managed communities, Bookchin is largely of the 'old' social school. He concurs with long standing criticisms of egoist individualism as 'petty bourgeois exotica'. But this 'exotica' was immensely dangerous, helping nurture American anarcho-capitalism. Even so, he considers that individual anarchism was largely bohemian, exerting most influence in art, fashion and personal behaviour rather than in politics. Today, however, lifestyle anarchism has dangerously penetrated the milieu of radical politics and tainted it with a self-centredness and 'trendy posturing' that erodes the very 'socialistic character of the libertarian tradition' (1996). Hakim Bey's ontological anarchism is singled out as particularly loathsome. Bookchin finds in Bey's 'narcissistic anarchism' a 'postmodernist withdrawal into individualistic "autonomy"', Foucauldian 'limit experiences', and neo-Situationist 'ecstasy' – all condemned as complicit in rendering the 'very word anarchism politically and socially harmless' (Bookchin 1996).

There are of course other readings of individual anarchism, as already discussed. Certainly Stirner's pushes to ostensible extremes the quest for the ego and the seeming dislocation of the social from conceptions of human nature. While some argue that Stirner does not reject 'social unities' *per se* but, rather, essentialized and universalized conceptions of the social, his tendency to see 'the people' simply as 'an artificial entity created by power' (Newman 2002, 235) would certainly not have won him many friends among the collectivists of the time. Nonetheless, as Tormey (2005, 403) points out, rather than a direct rejection of collectivism, Stirner instead cautions dissidents to 'continually review the terms and conditions of our interactions with others' so that we ensure our alliances with others are always 'reviewable, contingent, held open'. In promoting autonomous spaces for individual flourishing, and in cautioning against the seductions of universalism and essentialism, Stirner seemed to offer a 'critique of ideology *as such* ... a space without constraints, obligations, contracts, permanently binding rational or "universal" features' (Tormey 2005, 403).

Conclusion

It is in this 'space without constraints' that much contemporary anar-chist activity is occurring today. New anarchism, and post-ideological anarchism, still embrace specific political positions and promote spe-cific values. But in going 'beyond ideology' and 'beyond vanguardism', a strict ideological conformity to these values is not demanded. This is not the same as saying that no underpinning values exist, however. There remain core values and core political 'bottom lines'. These include a championing of autonomy, anti-authoritarianism and social freedom – and a continued determination to see the state as the prob-lem rather than the solution. Many of these core values are derived from old anarchism and made more relevant to the current political landscape by new anarchism.

Post-ideological anarchism, unencumbered by the ideological inhibi-tions of its forebears, exercises more choice over how it assembles its particularistic politics. In arguing that modern anarchism 'has long ... needed a major overhaul' Purkis and Bowen (1997, 3) draw hope from an emerging 21ˢᵗ century anarchism responsive to changed political conditions and 'firmly rooted in the here and now'. More recently they have identified the beginnings of just such an overhaul, one informed by poststructuralist insights and a broader conceptualization of power (2004, 21). In a similar vein, Milstein (2004) states:

> ... a 'new anarchism' seemed to have been birthed amid the cold rain and toxic fog that greeted the November 1999 World Trade Organisation protest. Yet rather than the bastard child of an emer-gent social movement, this radical politics of resistance and recon-struction had been transforming itself for decades. Seattle's direct action only succeeded in making it visible again. Anarchism, for its part, supplied a compelling praxis for this historical moment.

We turn next to exploring the (re)birthing of the anarchist impulse in the anti-globalization movement. In the process we begin to uncover the elements of anarchism's 'compelling praxis' that animates 21ˢᵗ century dissent.

2
Movements of Anti-Globalization

In 1994 the Zapatista's Mexican insurrection helped name both the enemy – neo-liberal globalization – and the sentiment against it – enough! Taking this cue, the anti-globalization movement (AGM) framed its dissent in this spirit and in these terms. The Zapatistas' compelling act of rebellion, and their clear illustration of neo-liberalism's dire impacts on the marginalized, helped provide oppositional politics a focus and an impetus. Contemporary dissent is driven by antagonism to a globalization identified as culprit in widespread social and environmental ruin. As capitalism's latest stage, this globalization is understood as neo-liberalist and corporate, and particularly adept at sharpening the inequalities that have always defined capitalism. Hence the development of an oppositional movement that is *against* neo-liberal globalization and *for* global justice. This chapter does not consider the debates about the status of globalization as a new phenomenon, nor the degree to which it signals a radically deregulated trading environment in which states are made irrelevant. These are of course vital debates that yield important information about the character of the current socio-economic climate (Hirst & Thompson 1996; Weiss 1998; Scholte 2000; Giddens 2002). There is nonetheless general agreement that globalization describes a substantially transformed economic, political and cultural landscape across the globe. The AGM has made very clear where it sits in these debates: it is unequivocally against a globalization directed by the powerful for the benefit of the few. It names as its 'enemy' a neo-liberal version that transfers power to corporations who then direct increasingly emasculated and compliant states to do their bidding.

Globalization has pressed radicals to reconsider any traditional reliance on the state for social renewal. Anarchists had always been dismissive of the state's transformative capacity, seeing it as the problem rather than the solution. But many on the left had relied on it to at least tame the excesses of capitalism. Scepticism of the state has penetrated a disenchanted left many of whom see in globalization the crippling of any kind of redemptive state. Those radicals who had put their faith in socialism's 'enabling' state, and the more recent social democratic state, were left feeling particularly bereft. They admitted, even if reluctantly, that corporate globalization signals a state deeply compromised in its capacity for social justice. Furthermore, the state is not only emasculated in the face of corporate power, but also complicit in its very transfer. The state's dogged determination to utilize its vast institutional (and military) resources to underwrite neo-liberal globalization, leaves radicals little choice but to stridently oppose it. Moreover, what were previously considered reasonable social demo-cratic demands are now increasingly viewed as radical – leaving dis-senters little choice but to occupy an oppositional space not necessarily of their choosing. As Hari (2002, 22) points out: 'it now seems that the traditional social democratic goals of a mixed economy and sharing the wealth of capitalism with the poor have become so radical (and so contrary to the agenda of the IMF and World Bank) that they can only now be pursued outside parliaments'. This widespread loss of faith in state capacity and in traditional political agency, has helped open up a new oppositional landscape nurtured by anarchist ideas.

We use the name anti-globalization movement to describe this dissent, but for practical rather than more pointed reasons. The term anti-globalization simply situates contemporary dissent within the broader politics of globalization as well the broader political climate that this encapsulates. The issue of naming is nonetheless a con-tentious one that raises important concerns. The chapter thus begins by considering how best to identify, and name, what the movement is for as well as against. The chapter then traces the development of the anti-globalization movement (AGM), distinguishing it from the new social movements that preceded it but also noting the considerable continuity between them. The AGM's key elements are next identified and discussed. The AGM is distinct in many ways, but it is not an entirely new movement. It can best be conceptualized as a movement that draws from a variety of oppositional strands – old and new – and then reweaves them into a distinctive politics buttressed by the new technologies made available to it. Many of the constituent groups com-prising the AGM come together at the annual World Social Forum. This forum, or movement, provides an open space for the coalescing of the disparate, diverse and autonomous voices united in their opposi-tion to neo-liberal globalization. We explore this social forum process and its links with the AGM. Throughout we identify and discuss these movements' anarchical impulses.

Naming the movement

Burgmann observes that the term globalization needs a 'qualifying adjective' if it is to make any sense (2003, 244). Applying adjectives such as 'neo-liberal', 'corporate' or 'free market' to globalization gives a clearer sense of what it is. Globalization, as the latest stage of capital-ism, is after all a very sweeping term that refers to a broad range of activities, processes and politics. Opposition to globalization demands similar qualifiers if it is to identify what it opposes and the sources of its discontent. In 'naming the enemy' as corporate globalization Starr (2000) helped identify its opposition as the anti-corporate move-ment. She contends that globalization has invested corporations with immense power, and that they proceed to use this power to wreak social, economic, political and environmental ruin on many commun-ities across the globe. Many different kinds of movements are thus 'naming corporations as their enemies' and 'working to envision alter-native political economies that meet their goals without empowering corporations' (2000, xi). In coining the phrase 'naming the enemy' there is a resolute assertion that 'an enemy exists and that it is recog-nizable' (2000, 1). This enemy is most starkly visible in the structures and processes of contemporary globalization. But this enemy also goes beyond structuralism, to feed off sources of power that are more complex and multiple:

> Structure works not only through political economy, but also through culture, basic and sophisticated social institutions, technology, and political theory ... a racialised capitalist world system is more than a set of institutions, it is clearly also a Foucauldian 'episteme', an entire (and culturalised) system of power and knowledge. No such recogni-tions of complexity dissolve the enemy or its enmity (2000, 5).

It is the relentless mining of most resources – physical and cultural – in the rush for profits that most disturbs Starr, and most anti-globaliza-tion dissenters.

Klein (2001) too identifies transnational corporations as the drivers of neo-liberal globalization. These corporations are emboldened by deregulation policies and empowered by technologies that significantly extend their temporal and spatial reach. She identifies the processes of branding and logos as the way in which corporations consolidate their power, maximizing their profits with worldwide sourcing and brand sovereignty. Enabled by wins in trade liberalization and labour-law reforms, these corporations are better able to source their labour and raw materials – and expend their externalities – in (usually poorer) regions that provide the best comparative advantages. Large trans-national corporations such as Nike and Microsoft not only manufacture products, but also maximize their profits through the marketing of the brands and logos that secure their products:

> What these companies produced primarily were not things ... but *images* of their brands. Their real work lay not in manufacturing but in marketing. This formula ... has proved enormously profitable, and its success has companies competing in a race towards weight-lessness: whoever owns the least, has the fewest employees on the payroll and produces the most powerful images, as opposed to products, wins the race (Klein 2001, 4).

According to Klein, the voraciousness of consumption is driven by 'branding'. The worldwide sourcing on which branding's profits rest, exploits the world's most marginalized peoples as well as the ecological systems on which they depend. In wreaking multiple havocs on multiple communities, corporate globalization also inspires multiple resistances. This oppositional diversity – indigenous movements, unionists, farmers, environmentalists and feminists – coalesces in the AGM.

While this oppositional diversity is clearly against a neo-liberal globalization, or capitalism, driven by corporations, and while it is important to focus its resistance as such, there are other important reasons for thinking carefully about its name. The anti-globalization movement implies an oppositional monolith when in fact is it is composed of, and prides itself on, diversity. The increasing use of the name 'movement of movements' captures well the fact of a movement that is made up of many parts. The movement is also concerned to project a positive image, so that its own 'brand' articulates well what it is *for*, as well as *against*. It is for this reason that growing sections of the movement prefer the name global justice movement. September 11 also left its mark on the movement's identity. Anti-globalization protestors were concerned to more clearly distinguish their movement from a terrorist anti-globalization that the public might confuse them with.

Global justice thus re-branded the movement in a more positive light and refocused its oppositional stance. As Monbiot contends, 'what we need now is to move from being an opposition movement to being a proposition movement' (in Hari 2002, 22).

Finally, it is important to consider the degree to which the movement, and the World Social Forum after it, can or should be understood primarily as an anti-capitalist movement. Critics of the movement might find the anti-capitalist label useful in a pejorative sense. However, while there is a very strong anti-capitalist element within the movement, anti-globalization protestors are not all necessarily anti-capitalist, or at least conceive their anti-capitalism very differently. This is not unexpected in a movement that champions its diversity, a diversity that ensures it multiple political views. A significant danger of the 'anti-capitalist' label is that in creating an oppositional singularity, it risks the movement's compositional autonomy. Even so, its multiple political views take on a decidedly anti-capitalist flavour. Most movement observers differentiate between the reformist and radical elements of the movement, a distinction that has applied to most social movements since they have always contained a spectrum that ranged from the reformist to the radical. But the anti-capitalist impulse that runs through the movement is a very important and defining one, and is, in any case, the subject of this book. The 'new' anti-capitalism carved out by the AGM is also important in another way. It helps define a new oppositional landscape that 'resume[s] a battle now largely abandoned by established political parties of 'the left' (Burgmann 2003, 276). It is also a new oppositional landscape increasingly infused with anarchist ideas.

Old, new and newest social movements

All social movements are induced by their specific political climates and powered by the resources available to them. As products of explicit historical conjunctures and political configurations, social movements set out to make changes in the world that best reflect their principles and priorities. We have already noted how the New Left helped shape and propel the new social movements of the late 1960s. Here we saw the emergence of a new set of actors that conceptualized a much broader operation of domination and attempted to transcend the restrictions that the old left placed on them. Informed by some of the newer discourses of postmodernism and poststructuralism, these new social movements also moved beyond the New Left to forge a much more particularistic politics of identity. Women, indigenes, gays, environmentalists, peace activists and a broad range of others, moved centre stage to drive the complaints of the new social movements.

The old left sprang from an industrial age where worker rights and privileges were limited. In this context, their focus on labour relations and working conditions, in seeking the best deal possible for the working class against the demands of the bosses, made much sense. As Touraine (1974) points out industrial society problematized labour relations while 'postindustrial' society problematizes a broader range of human experiences, many of which have been taken up by the new social movements. Labour remained a problem but it was now one among many. Unlike the old social movements that situated labour/unionist struggles as the main agency of social change, the new social movements reconceptualized and diversified social struggle. The labour movement was not cut off altogether, however, continuing to play an important role in the complementary struggles of the new movements (see Burgmann 2003, 16–17). But in disentangling social change strategy from its institutional focus, and in eschewing links to political parties and the route of statist power in general, the new social movements problematized the role of the 'old' labour movement within their ranks. Even so, hierarchical and leader-focused organizational structures were retained, as demonstrated in the organizational character of the many non-government organizations (NGOs) that ensued. In promoting an invigorated civil society, and new postmaterialist values that transcended the old left's distributive focus, the new movements expressed a certain disdain for the materially-focused and 'unreconstructed' generation of old left activists (see Touraine 1974; Habermas 1981; Offe 1985; Beck 1994, 1996).

Largely anti-systemic, and embracing new values, the new social movements challenged the broader cultural and materialist underpinnings of capitalist society. But the highly individualized identity politics that drove it proved both its strength and its weakness. For some the 'politics of difference' impelled the movements to lose much of their 'critical, collectivist, confrontationist and campaigning impetus' (Burgmann 2003, 24). Others lamented the movements' championing of a 'hyper-individualism' that played into the very hands of their capitalist opponents who were promoting just that (Frankel in Burgmann 2003, 26). The impetus of the new social movement stalled in the 1980s for a variety of reasons. One important one was the absorption of many of their claims in government policy by social democratic, and even some conservative, governments. NGOs increasingly lobbied governments to incorporate their concerns into government policy. While many of the new movements did not seek direct statist power, and direct links between parties and the movements were largely eschewed, the central strategy of demanding institutional change remained. To this degree, the new social movements' successes in placing their concerns on the mainstream agenda paradoxically spelled their 'demise': if their concerns were embraced, their *raison d'etre* diminished. At the very least, many of them were forced to reconsider strategy in the face of these and other broader political developments, including the politics of globalization. Many of them turned their attention to supporting the pro-democracy movements of Eastern Europe in the 1980s. And as social democracies' labourist and welfare policies were increasingly pressured by the demands of neo-liberal globalization, attention turned to the mounting of a new line of defence.

This new line of defence consolidated around the theme of anti-globalization. But it is a dissent still composed of many elements of the old, albeit compelling in its capacity to link together many new, old and disparate anti-globalization pieces. Continuities between old, new and newer (anti-globalization) movements remain important. Rather than old social movements withering away altogether and being replaced by new ones, there is instead a transformative continuity whereby old movements, like the labour movements, change to accommodate the new times (see Cohen & Rai 2000; Burgmann 2003). Control of labour relations, after all, underpins capitalism's logic, so it is no surprise that labour movements continue to contribute important pieces to the AGM's oppositional mosaic. But these labour movements interact in a considerably refashioned oppositional landscape. Rather than leading an old left charge, they instead form part of a multiplicity of dissent or oppositional network. As Stahre contends, '[c]ooperation in the form of networks of groups and umbrella organizations seems generally to be an innovation of contemporary social movements compared to the movements of the 1960s' (2004, 77).

As a spirited progeny of the new social movements, the AGM has reconfigured its dissent in response to altered political conditions, past lessons learned, and the technological opportunities afforded it. The AGM sought to weave its different elements into a coherent framework. It did this not by eschewing fragmentation – diversity was after all built on it – but by gathering these fragments into an oppositional unity that simultaneously empowered the movement as well as its constituent parts. The movement's innovation was thus its network structure – a structure assisted and shaped by the new technologies available to it. Networks enable the establishment of links and alliances between the disparate parts, retaining oppositional autonomy within a network of oppositional unity – a phenomenon explored in detail in Chapter 3. This replaces the plethora of seemingly separate battles encapsulated by the new social movements. Strategically this networked 'fragmentation' also acted to protect the movement against its opponents: it becomes difficult to attack the movement's core if it does not have one. Of course, the identification of *an* enemy around which to pivot the movement raises its own risks or 'disempowering effects', as Starr (2000, 29) herself notes. We have already observed how, in training its sites on class politics, Marxism was charged with overlooking many other forms of oppression such as those based on race and gender. Furthermore, the idea of an enemy can imbue globalization itself with more power that it ordinarily possesses. Hirst and Thompson (1996) contend, if in a different context, that a totalizing conception of globalization infers a completeness and power that overlooks its fragmentation and fragility. But it is at the juncture of this fragmentation and fragility that the AGM inserts itself. We turn now to exploring its evolution and key features.

The development of the AGM

The AGM has targeted the central forums of global governance – among them the World Trade Organization (WTO), World Economic Forum (WEF), G7 and G8, International Monetary Fund (IMF), and the World Bank (WB). These organizations make major global political and economic decisions at annual summits across the globe – decisions that effect the operation of globalization worldwide. They are thus seen as the representatives and drivers of globalization. Protest against the November/December 1999 WTO meeting in Seattle constituted the AGM's first major action, and introduced it to the world as such. The 'Battle for Seattle' involved an estimated 100,000 protestors with over 500 arrests. The Seattle action was notable for its compositional diversity – environmentalists, unionists (the largest grouping), anarchists, peace activists, church groups, international socialists, farmers, and many others. The shared complaint was against a neo-liberal globalization that degraded the environment, created unemployment, worsened working conditions, threatened farming livelihoods, destroyed communities, exploited indigenous peoples, abused the third world, produced the conditions for war and, overall, undermined equality, justice and democracy. The September 2000 meeting of the WB and IMF in Prague attracted a similar – if some-what smaller – protest. Also in September 2000, the S11 protest in Melbourne against the Asia-Pacific Economic Summit of the WEF, drew over 20,000 protestors for what was to be an event marred by charges of police brutality. In April 2001, the Summit of the Americas in Quebec City drew a crowd estimated to be even bigger than that of Seattle. The subsequent July 2001 protest against the G8 summit in Genoa proved a pivotal moment in AGM actions. With an estimated 200,000–300,000 protestors – twice or three times the number of Seattle – the Genoa protest claimed its first casualty, 23-year-old student Carlo Giuliani. Genoa was marked by claims and counter-claims of police brutality and protestor violence, property damage estimated at over $50 million, and negative media coverage overall. Regardless, the subsequent March 2002 action in Barcelona attracted over 300,000 protestors.

The most recent anti-globalization event, at time of writing, was that against the 31st G8 summit held from 6 July to 8 July 2005 at the Gleneagles Hotel in Perthshire, Scotland. Its agenda items included the 'problem' of African poverty, global aid and debt strategies, counter-terrorism measures, reform in the Middle East reform and climate change. Two strands of anti-globalization opposition coalesced during the Scottish event. One, Bob Geldolf's *Making Poverty History* Live8 campaign collected together an umbrella group of church groups, char-ities, trade unionists, celebrities and general protest groups to demand debt relief and improved aid and trade for Africa. Attracting a crowd of over 200,000, the pre-summit event was claimed to be Scotland's biggest demonstration (BBC News 2005). Not necessarily aligned with the AGM – and indeed incurring considerable criticism from it for upstaging its own anti-summit event and 'celebritizing' protest in general – the *Making Poverty History* demonstration attracted extensive, and largely positive, media attention:

> LIVE8 has finally met G8 with a challenge to the world's leaders to sink their differences for the sake of a comprehensive deal to allevi-ate African poverty, and a warning from Bob Geldof that they were playing 'the highest poker' with people's lives...The Live8 rock singers had no time for the violent protesters blocking roads, smash-ing windows and attacking police, and contrasted them with the 'positive' [Make Poverty History] march on Edinburgh at the weekend (The Australian 2005, 8).

The second main strand, was the anti-globalization event 'proper': the G8 Alternatives protest to run parallel with its formal summit. The Dissent Network set itself the task of co-ordinating a variety of radical resistance actions to the summit, including blocking access to the Gleneagles venue itself (BBC News 2005). It attracted significantly fewer numbers than at the *Make Poverty History* march and less positive press. But much media attention was trained on what was expected to be anarchical chaos. Leading with 'Anarchists put chaos on agenda for summit', *The Scotsman*, for example, wrote:

> Fears that extremist groups will head to Scotland for the G8 meeting this summer are rising ... [Security measures are] all part of a bid to prevent the kind of scenes that blighted the G8 summit in Genoa, Italy four years ago, when hundreds of protestors were injured and one killed following horrific scenes of rioting ...it's that kind of situation that police chiefs and security experts fear may happen here ... [driven by] Internationale of Anarchist Federations – a group aimed at building up international anarchist structures with links to groups across (Dick 2005).

As it turned out, there were a number of clashes with police as several thousand protestors converged on the Gleneagles site and attempted to scale the security cordon that surrounded it. By all agreement, however, these clashes were inconsequential compared to those in Genoa and Seattle, and other protests.

Actions that captured security concerns included the Carnival of Full Enjoyment, an anarchical carnivalesque event employing a Reclaim the Streets ethos of 'roving' anti-capitalist street parties. As one participant observed, this oppositional event was: 'a celebration, a creative and positive thing ... Music, dancing, drumming, singing, making new friends, and all in solidarity against the brutal exploitation of the capitalist system and the (fresh in our minds) brutal tactics of the police' (WOMBLES News 2005, 4). Other notable anarchical events were the erection of an alternative eco-village at Stirling, showcasing non-hierarchical and self-governing communities in action. Here, as an involved activist explained, the thousands of protesters heading for Gleneagles 'are camping beside the River Forth in Stirling...[and] with the co-operation of the local authorities, they are living according to alternative social and ecological principles...[T]he first principle of what they call Hori-zone is that there is no hierarchy and no leader. The people govern themselves by collective agreement' (in Mitchell

2005, 1). But anti-summit protests, and the outcomes of the summit itself, were significantly overshadowed by the London underground bombings of 7 July. All – protestors and summit officials alike – were humbled by the bombing's carnage.

Even before G8 2005, however, there had been considerable reflection on the effectiveness of the AGM strategy of summit targeting. The diminished numbers at the Gleneagles protest echoed this. Several factors contributed to this strategic reflection, including the post-September 11 environment, a growing anti-war movement to which many in the AGM had turned their attention, as well as a general rethink of the continued efficacy of large summit protests. But the protest tide was turning even before September 11. The July 2001 protest against the G8 in Genoa represented its own turning point. Long held tensions in the movement between those who advocated violence – usually to property, mainly symbolic and frequently led by the Black Bloc – and those who opposed it, bubbled to the surface after the tragedy of Genoa (see Albertani 2002). The AGM could no longer ignore how its own strategies were being countermanded and infiltrated by police and other security forces, and how the media increasingly reported and published images of a violence that distorted the image of the AGM as a whole. This is despite the fact that one of the defining features of the movement is its general rejection, 'even among its most militant elements, of either armed revolutionary struggle or terrorism (along the lines of the Red Brigades or Wethermen just a generation ago)' (Panitch 2002, 13). Others reinforce this view, claiming that it is the movement's *lack*, rather than *presence* of, violence that disturbs governments. According to Graeber (2002, 66) 'what really disturbs the powers-that-be is not the "violence" of the movement but its relative lack of it'; this is because 'governments simply do not know how to deal with an openly revolutionary movement that refuses to fall into familiar patterns of armed resistance'. Fair play or otherwise, the movement was stung by such reports of violence and its image dented. As a result it was left vulnerable.

For some, the wars in Afghanistan and Iraq represented a 'godsend' for the movement, stemming its seeming decline and resharpening its focus. The anti-war demonstrations that involved millions and that were staged in all corners of the world on 15 February 2003 are generally considered one of the peace and AGM's most spectacular mobilizations. While anti-war and anti-globalization remain distinct movements, there is considerable interchange of its participants, especially in a post-Iraq war climate. Many believe that further collaboration on an even broader front would benefit both sides. Epstein (2003, 116) argues that if the anti-war movement is to 'gain strength and momentum' it 'needs to link up with the broader anti-globalization movement'; conversely, the anti-globalization movement would benefit from closer links to 'its labor and environmental segments' especially by uniting under the broader 'critique of commodity fetishism' (Epstein 2003, 116).

Understanding the AGM

Despite different organizational forms, most social movements share some defining features. Following Touraine, Castells (2004, 74) highlights three of these. The first is the movement's *identity*, that is, how it sees itself and 'on behalf of whom it speaks'. Second, since movements stand in opposition to something, each has an *adversary*, or 'enemy', that it must clearly identify. Finally movements are motivated by a *social vision* – a transformational goal to which they aspire and towards which their opposition is directed. While a movement made up of many other movements, the AGM still fits these criteria well. It has a clear identity, a specifically named adversary and a shared vision of global justice. Many other theorists have elaborated on these features – albeit with different degrees of emphasis. In this section we utilize, and build on, the insights of Buttel (2003), who identifies some key structural characteristics of the AGM that provides a useful way of organizing and discussing them. But we go considerably further, illuminating throughout the movement's anarchical temperament.

First, the movement is clearly a global movement, or what others have called a 'global dialogue of the oppressed' (Tormey 2004, 62). While the movement's northern arm is clearly visible, its southern one is equally significant. Buttel points out that, as of 2000, anti-globalization protests took place in approximately 75 cities on six continents around the globe (2003, 99). Even a very quick purview of anti-globalization actions in the past few years demonstrates their global spread. These include the landless movement of Brazil, the protest villages of Thailand, the mobilization against Monsanto in India, resistance against the 'oil occupation' of the Niger delta, Kenyan resistance to World Bank policies, Argentina's movement of unemployed workers, anti-eviction actions in Johannesburg, the Zapatista insurgencies in Mexico, to, of course, the north's Seattles, Pragues, Melbournes, Quebec Cities, Genoas and Edinburghs (see Notes from Nowhere 2003). The First Intercontinental *Encuentro* (meeting) for Humanity and Against Neo-liberalism, held in the depths of the Chiapas jungles in Mexico, attracted thousands of people from five different continents (Notes from Nowhere 2003, 126–7). Even so, in compiling its list of participants, the Notes form Nowhere collective observed that 'there were many places we couldn't reach, where barriers of language, culture and distance prevented us hearing the voices of those directly involved' especially in the South' (2003, 15).

Burgmann (2003, 264) labels the AGM 'the new internationalism', highlighting the plethora of community-based actions throughout the globe. In *Naming the Enemy: Anti-corporate movements confront globalization*, Starr (2000) details a number of these local, national and global movements, and, importantly, how they intersect. Kingsnorth's (2003) 'journey to the heart of the global resistance movement' directly acknowledges the 'global coalition of millions' that it constitutes. Nonetheless, reflecting superior access to resources and a media willing to report them, northern actions captured considerably more attention. But this should not deflect from the fact that against the expectation of a north dominated movement, 'the lion's share of protests have actually occurred in the South' particularly 'in Bolivia, Argentina, Thailand, India, Brazil and Indonesia' (Buttel 2003, 99). And it is southern movements such as Mexico's Zapatistas that have contributed a core element of the global movement's inspiration, helping shape its ethos and strategy, especially in the north.

This global phenomenon goes beyond reach, however. It also highlights considerable connectivity between northern and southern actions, with novel alliances emerging between the two. Given globalization's transformation of capitalism into a transnational force, so too has its opposition been 'transnationalized', or globalized. While, as della Porta and Kriesi (1999, 4–5) point out, 'nation states are still the principal actors in international relations, and the national political context continues to constitute a crucial filter which conditions the impact of international change on domestic politics', globalization has necessarily altered the nexus between the local, national, international and global, and with it, the global character of dissent:

> [I]n the globalising world issues emerge which transcend national frontiers ... The challenge for the (national and transnational) social movements is to develop the organisational structure, action repertoires and strategies necessary to address a multi-level system of governance – a system which ... provides them with new constraints but also with new opportunities (della Porta & Kriesi 1999, 21).

A constraint for some is an opportunity for others. In a transnational environment, the necessity of bypassing the state to effect change, and the necessary creation of a networked movement structure suitable for these times, becomes the movement's anarchical opportunity. This permits more direct and less mediated forms of opposition, ones that targets culprits such as corporations directly through autonomous and localized actions. For Castells, this reverses the well known slogan, 'think global, act local':

> Because power increasingly functions in global networks, largely bypassing the institutions of the nation-state, movements are faced with the need to match the global reach of the powers that be with their own global impact on the media, through symbolic actions ... Reversing the popular motto of twenty-five years ago, social movements must think local (relating to their own concerns and identity) and act global – at the level where it really matters today (2001, 142).

We should avoid over-generalization, however. The character of the south's oppositional politics is considerably different. Because of the very different political make up of some southern states, and very different existing social and authority structures, opposition often targets heads of state and senior politicians and bureaucrats more directly (see Buttel 2003, 99). Furthermore, southern movements are also significantly under-resourced relative to their northern counterparts and thus limited in the protest forms they can adopt. While the networked character of the movement is fuelled by the new technologies, many in the south do not have access to such technologies, or to the range of other available tools, such as mobile phones, that power its northern arm. It is for these reasons that Buttel claims the northern arm of the movement as the most influential (2003, 99). Yet enhanced network capacity powered by new technologies remains an important part of the movement story, even for the south. These technologies have helped develop a 'global (anti-capitalist) village' enhancing 'the *capacity* of activists to organize and cross-pollinate'; to undertake a genuine global dialogue; to coordinate their activities; and for the disparate protest groups, in realizing that they were not alone, to offer each other visible encouragement and support (Tormey 2004, 63–9). The important role of technology is detailed in Chapter 3.

The second structural characteristic of the AGM is its diversity. The movement contains not only the 'usual suspects' – environmentalists, anarchists, socialists and peace activists – but also a broad range of others: unionists, church groups, the unemployed, farmers, consumer groups and so on. This was a novel collaboration between different segments of social movements that were often oppositional in the past. Anti-globalization protest saw these groups come together as oppositional comrades rather than foes, united against the common enemy of neo-liberal globalization. There was a preparedness to put differences aside in the interests of the bigger battle – even as significant tensions remained. This diversity was showcased in the Seattle protest. Unionists were probably the largest grouping, outnumbering environmental activists and other groups of 'traditional' movement activists. The new oppositional unity of two previous foes was exemplified in the slogan: 'Teamsters and Turtles – together at last'. But this 'unity-in-diversity' went further. As one Seattle observer noted: 'I saw anarchists marching uneasily alongside men carrying banners of Stalin, and I met Christian human rights groups who were profoundly uncomfortable at being in the company of either' (Hari 2002, 20). This diversity is also captured by the movement motto of *One No, Many Yesses:* a resounding 'no' to neo-liberal globalization, and equally resounding 'yesses' to the expressive diversity of this dissent. Exporing the AGM's diverse and pluralist movement character, Kingsnorth (2003) documents its expression in five different continents, concluding that rather than socialism, capitalism or any other ism animating the movement, it was instead 'united in what it opposes, and deliberately diverse in what it wants'.

Others label this diversity a 'unity of many determinations', highlighting the 'unprecedented alliances' between oppositional forces one would have 'expected to only be interested in particularistic, identity-based organizing' (Starr 2000, 158). These 'many determinations' include a broad array of autonomous and often allianced players: opponents of structural adjustment policies, welfare reform alliances, peace and human rights supporters, land reform movements, the labour movement, anarchists, cyberpunks and consumer groups (Starr 2000, 158–61). Similarly, Klein (2001) calls the movement 'coalitions of coalitions' to identify a unifying protest core made up of many parts. These many parts reflect the various ways neo-liberal policies devastate many communities throughout the world. Starhawk – a noted eco-feminist – has distilled from this diversity, nine key commonalities or 'deep principles and imperatives' (2003, 237–41). While not all anti-globalization protestors would choose Starhawk's wording, nor support every principle as articulated, together these enumerated principles reflect the diverse demands of the movement:

We must protect the viability of the life-sustaining systems of the planet, which are everywhere under attack. A realm of the sacred exists, of things too precious to be commodified, and must be respected. Communities must control their own resources and destinies. The rights and heritages of indigenous communities must be acknowledged and respected. Enterprises must be rooted in communities and be responsible to communities and to future generations. Opportunity for human beings to meet their needs and fulfil their dreams and aspirations should be open to all. Labour deserves just compensation, security and dignity. The human community has a collective responsibility to assure the basic means of life, growth, and development for all its members. Democracy means that all people have a voice in the decisions that affect them, including economic decisions.

While the diversity of the AGM is an important characteristic, and while many of the groups that comprise it have specific priorities and goals, at the same time the AGM has a 'strong collective identity and very strong political culture' that not only opposes corporate power but also rejects 'the consumerism and the dominant ethos of modern capitalism' (Buttel 2003, 101–2). In short, while unequivocally diverse, the AGM still has a shared worldview.

Diversity can come at a cost however. In a political culture that values unity, the AGM's diversity provides opportunities for its critics to disparage it and security forces to undermine it. This was reflected in the internal disputes on property violence between activist groups prior to the Genoa event. The role of violence has manifested controversially in Italy, reflecting its historical experiences with autonomist actions and various insurrections against the state. Many claim that Italian security forces took advantage of this political history to destabilize the movement by associating it with this 'dark stain' in Italy's oppositional past. Starhawk asserts, for example, that the Italian police either infiltrated or provoked elements within the movement, encouraging direct property damage that they could then 'heroically' contain (2003, 101–32). Others claim that 'there were cops in ski-masks leading the more excitable and native among Genoa's young bloods on attacks on corner shops, bus stops, and post offices' (Moore 2003, 368). Since the protestors' first goal was to penetrate the fortified police lines barring entrance to the summit, the Italian anti-globalization group *Tute Bianche* – appropriately 'uniformed' in protective clothing and 'armed' with plastic shields – took charge of penetrating the police cordon. This stand-off provided police with their retaliatory justification, unleashing the barrage of tear gas, water cannon blasts and vigorous physical resistance that ensued (see Notes from Nowhere 2003, 356–70).

Aside from the 'degeneration' of the Genoa action, the protest also exposed some of the tactical fault lines intrinsic to the movement's organizational diversity. In particular it highlighted some of the tensions between the movement's reformist and more radical elements, tensions that went beyond the issue of violence. The tactical and political winding of Genoa, and the recognition that a more reformist, united and less strident movement is better able to capture sympathetic media coverage, has prompted the movement's considerable reflection, as we already saw. How the media presents the movement is critical to its sustenance, with the dissemination of the movement's principles increasingly dependent on this fickle media. Participants nonetheless caution against panic and the imposition of 'a single perspective on the whole movement ... for whom diversity is a central value and goal' (Viejo 2003, 371–2). Furthermore, if diversity comes at a price, it is also a tactical strength. A movement activist contends that the movement's robustness relies on the dynamic between its radical and moderate arms (Viejo 2003, 373). The tactics of the Black Bloc radicals, for example, 'reawoke the world's political imagination' and captured necessary public attention, while the moderates protected the radicals against arrest and retaliation. In a dialectical interplay, the moderates contain the radicals' extremes while the radicals compensate for the moderates' 'timidity'. According to this view, this balancing act ensures the movement's vibrancy and diversity as well as its strategic unpredictability – critical to its protection and survival.

The third structural characteristic of the AGM clearly marks out its anarchist impulses. Here we identify the movement's networked, leader-less and non-hierarchical structure – a form assisted by the new technologies. Many observers locate these acephalous characteristics as central to the movement's character (see Buttel 2003, 100). A movement activist highlights their significance:

> The strength of this movement ... has always been its leaderless fluidity, its constantly changing strategy, its unpredictable tactics and targets. This is why the authorities [until Genoa] have found it so hard to get a handle on what we were up to – we weren't following patterns or playing by any discernable rules (Moore 2003, 369).

Klein describes the AGM's organizational form as protest 'hubs' incorporating a diversity of autonomous 'spokes'. But rather than signalling fragmentation, this organizational form instead reveals a political and strategic strength. This is because it stands as a participative contrast to the hierarchical forms adopted by the institutions and corporations the AGM opposes. This radical democratic form ensures the unity between means and ends that anarchism demands. This 'model of *laissez-faire* organizing' is thus not only 'extraordinarily difficult to control' but also 'responds to corporate concentration with a maze of fragmentation, to globalization with its own kind of localization, to power consolidation with radical power dispersal' (Burgmann 2003, 298). Starr agrees that the movement's strength lies in its constellation of non-hierarchical alliances, in the 'unprecedented alliances' forged by 'the very people who were expected only [to] be interested in particularistic, identify-based organizing' (2000, 158). And Graeber insists that:

> ... the general anarchistic inspiration of the [AGM] movement ... is less about seizing state power than about exposing, delegitimizing and dismantling mechanisms of rule while winning ever-larger spaces of autonomy from it ... [and all in a] general atmosphere of peace (2002, 69).

Acephalous political structures have always attracted considerable derision, however – a derision that underpins the general dismissal of anarchism as an impractical political ideology. This criticism is both internal and external to the movement. Internally, participants observe that an emphasis on non-hierarchical organizational forms, consensus decision making and autonomous protest hubs, can delay the speed with which globalization can be challenged and its damage stemmed. Externally, the movement is exposed to ridicule from the mainstream – from a range of political, corporate and media quarters that value leadership, hierarchy and centralization, and that compellingly argue that to do otherwise is to remain hopelessly utopian. Epstein notes the undermining of some of the 1960s movements because of 'structural and ideological rigidities associated with insistence on consensus decision making and reluctance to acknowledge the existence of leadership within the movement' (in Hari 2002, 22). The AGM's largely non-hierarchical, networked organizational form is thus viewed both as a fundamental strength *and* a fundamental weakness, both strategically and ideologically. The movement's anarchist thrust champions it, while its more moderate arm problematizes it. But it remains central to the movement's anarchical temperament as a whole.

A final characteristic highlights the disproportionately higher numbers of young people in the movement, a phenomenon clearly visible during protest events and generally noted by most movement observers. Epstein (2001) remarks that many activists in the United States are in their teens or twenties. High rates of youth participation has often characterized protest movements, especially in the north, and the new social movements were animated by large numbers of young, well-educated students. Youth protest is often a reaction to the politics and priorities of the previous generation. Many young people today have witnessed the disintegration of the Soviet bloc and the fall of the Berlin wall – as well as the consolidation of a triumphant neo-liberalism. Many would also have observed the ideological repositioning of their parents' generation as long held principles withered in the face of obstinate contradictions. In this post-ideological environment, younger generations are further removed from their parents' political heroes and bristle at expectations of ideological 'correctness'. But this does not signify a *de*politicization – rather a disavowal of sectarianism, and a reshaping of dissent to better respond to new challenges. The post-ideological anarchism that animates the AGM is just such a response, encapsulating well a radical politics suitable for its post-ideological times.

Many of the components and characteristics of the AGM also come together in social forums, and particularly the annual World Social Forum. The next section explores the WSF process, illustrating its links with the AGM and highlighting its anarchical impulses.

The World Social Forum

The World Social Forum (WSF) is an annual global gathering of social movements, NGOs and other civil society organizations opposed to neo-liberalism globalization, and motivated by the conviction that 'another world is possible'. As an open meeting outside the boundaries of the institutional state, the forum creates an autonomous space in which participants can share, debate and exchange ideas and proposals for the making of a better world. The forum process is also held nationally, with many countries hosting their own social forums annually or biennially. The first three WSFs were held in Porto Alegre, Brazil, before transferring to Mumbai, India, in 2004, and back to Porto Alegre in 2005. In 2006 the WSF was polycentric, held simultaneously in three regions: Africa, South America and Asia. Like the AGM's anti-summit strategy, the WSF is organized to run parallel to the WEF. Participants are from a wide range of backgrounds, herald from many different countries, hold an array of political views, and arrive armed with myriad proposals and strategies. But they share an aversion to neo-liberal globalization and the dispossession it imposes on their communities and their livelihoods. Like the AGM, this shared opposition is the common thread cohering them. The WSF, and forums in general, are also dynamic events that incorporate new developments and reconsider strategy in response to changing political climates. In the light of Iraq, the WSF has, as with the AGM, incorporated a resurgent anti-war, anti-imperialist current. And the increasing political successes of left parties in South America has buoyed many in the forum to press this institutional momentum harder.

Conceived of by a network of Brazilian and French unionists and activists with ties to the Workers Party (PT) and the Association for a Tobin Tax for the Aid of Citizens (ATTAC), the first meeting was seen as the beginning of a movement aimed specifically at 'creating proposals that would go beyond the growing protest actions against the neo-liberal model' (Hammond 2005, 31). Its original meeting place of Porto Alegre was carefully chosen, as were its subsequent locations. It was seen as important that its meetings take place in a southern location since the south not only bears the brunt of the neo-liberal onslaught but, importantly, is also the site of compelling resistance and alternatives to it. If the north's anti-globalization protests showcased *why* neo-liberalism should be challenged, the south would showcase *how* other worlds are made possible. Porto Alegre was not only in the south but this Brazilian location also represented hope. While a 'a city on the periphery' at the same time it was undergoing a process of democratic reform, led by a Left party committed to social renewal (Sader 2002, 91).

The first three years of the forum were overseen by an organizing committee composed primarily of Brazilian organizations, transferring to an Indian Organising Committee when the WSF was held in Mumbai in 2004. The 2005 forum returned to Porto Alegre and its Brazilian Organising Committee was comprised of 23 organizations divided into several working parties: Spaces, Solidary and Popular Economy, Environment and Sustainability, Culture, Translation, Communication, Mobilization and Free Software (World Social Forum 2004). The 6th polycentric WSF was decentralized and held in different southern regions in January 2006 in the city of Bamako in Mali, Africa and Caracas in Venezuela, Americas; and March 2006 Karachi, in Pakistan, Asia. Each WSF is subject to a charter of principles originally drafted in 2001 and approved by the WSF International Council (Patomaki &

Teivainen 2004, 145). Some of the charter's key points and principles include:

> [The WSF is an] open meeting place for reflective thinking, democratic debate of ideas, formulation of proposals, free exchange of experiences and interlinking for effective action... [that] respect[s] universal human rights, and those of all citizens – men and women – of all nations and the environment and will rest on democratic international systems and institutions at the service of social justice, equality and the sovereignty of peoples... The participants in the Forum shall not be called on to take decisions as a body, whether by vote or acclamation, on declarations or proposals for action that would commit all, or the majority... (World Social Forum 2002b)

Central to these principles is a non-hierarchical, horizontal and participatory organizational structure that extols self-determination, social justice and networked, autonomous resistance. Its political diversity sits at the forum's heart. As the coming together of global civil society:

> The World Social Forum is ... characterized by plurality and diversity, is non-confessional, non-governmental and non-party. It proposes to facilitate decentralized coordination and networking among organizations engaged in concrete action towards building another world, at any level from the local to the international ... (World Social Forum 2002a)

Still, many participants come to the forum very focused on their own specific agendas and seek very specific outcomes from it. Globalization – and the strategies for containing it – is also understood very differently by participants, reflecting the forum's accommodation of a broad spectrum ranging from reformists at one end to revolutionaries at the other. The forum's charter is nonetheless clear on the acceptable shape, principles and operation of the global meeting: the WSF 'does not intend to be a body representing world civil society' and it is 'not a group nor an organization' (World Social Forum 2002a).

The links between the AGM and the WSF are clearly demonstrated in the autonomous, diverse and anti-systemic ethos of both movements, one of the reasons why the media usually conflates the WSF with the AGM. While sympathetic, many activists reject such a fusion, however. They are concerned to promote the WSF as significantly different in character to the AGM. Unlike the AGM, the claim is that the WSF is not about opposition, but proposition – on the imagining of alternatives and the creation of other worlds. Even so, these alternatives are just that – a plurality of alternatives that reflects the fact that the WSF does not speak with one voice, nor *represent* a 'membership'. Participants in the WSF generally advocate stronger democratic processes as a main way to undermine the inherently undemocratic nature of globalization. Democracy also becomes both the means and ends of the forum process. The forum's horizontal structure and participatory inclusivity is seen as prefiguring a global democratic future, underwritten by an autonomous and invigorated civil society. In this way the AGM and the WSF share much – both underpinning their claims for global justice by a plural and diverse politics of opposition. It leads long-standing activists such as Susan George to call the forums 'high points of the movement year [that] ought to reflect both our evolution and the best we are capable of' (2004, 42).

The WSF draws on 'two broad currents of activism' that together constitute its anti-globalization: 'the direct action movement that has mounted massive demonstrations against international summit meetings' and 'the emergent world-wide civil society, embodied mainly in the non-governmental organizations that have mushroomed throughout the world since the 1980s' (Hammond 2005, 31). Wallerstein largely concurs, contending that the 'characteristics of this new claimant for the role of anti- systemic movement are rather different from those of earlier attempts' – that is, of both the old and the new social movements :

> First of all, the WSF seeks to bring together all the previous types – Old Left, new movements, human-rights bodies, and other not easily falling into these categories – and includes groups organized in a strictly local, regional, national and transnational fashion. The basis of participation is a common objective – struggle against the social ills consequent on neo-liberalism – and a common respect for each other's immediate priorities. Importantly, the WSF seeks to bring together movements form the North and the South within a single framework (2002, 36–7)

The last point is significant in distinguishing the WSF from the AGM – or at least in demonstrating the additional step that the WSF takes. While the AGM is a global movement that comprises dissent from both the north and the south, its protests have tended to occur as singular events in singular locations. The WSF brings together the north and the south in a 'single framework' that represents a distinctive development in movement politics. But this 'single framework' is also the source of significant tension in the forum over the identity and strategy it should pursue. This tension is conceptualized as one between 'space' and 'movement', a tension that resonates anarchist sensibilities.

Space 'versus' movement

As with the AGM, all is not harmonious within the movement. An embrace of diversity and a shared aversion to globalization does not deter a vigorous internal contest on the best way forward for the forum. More recently, the questions of political direction, strategy and relationship to the state that are continually debated in the forum, have led to a sharper division. In a forum process that extols diversity and difference – and where for the south the stakes are particularly high – this is not unexpected. The debate between the forum as 'space' or as 'movement' encapsulates this tension. While they may not specifically describe it this way, the tension resonates a contest for the forum's anarchical heart – framed in terms of autonomy (space) and institutionalization (movement), and resonating old Marxist/anarchist debates about strategy. These tensions have been reinforced by the recent successes of socialist parties throughout South America, prompting a galvanization of the socialist charge. This was reflected in the 2006 Venezuelan chapter of the WSF, with increasing attention devoted to the 'other worlds' that socialist governments were indeed making possible in the Americas. Not all were convinced, however. As one 2006 forum participant noted, 'the whole question of how revolutionary governments can be made within the world system of capital and the internal contradictions of political traditions' remains of 'primary concern' (Shor 2006). This prompted a workers' movement leader to ponder whether any 'political party, no matter how left-wing they are, has been able to take political power without succumbing to the dynamics of electoralism and moving to the right' (in Shor 2006).

These tensions mirror the tussle for the forum's identity They raise the issue of how best to take the oppositional movement forward: whether the forum should cease being a 'talking shop' and begin 'taking power'; whether it should abandon its distaste of statism by claiming formal political power; and whether it should cease being a 'space' and become a political 'movement' fixed on institutional reform. As currently conceived, the WSF does not generate uniform proposals or political statements that represent all of those involved. To do so would, as we saw, go against its charter of principles which promote it as an open meeting place that upholds the independence of its participants. The WSF, in short, does not speak on anyone's behalf. The forum simply makes available this open meeting 'space' where groups and individuals intersect and interact 'horizontally', extolling the virtues of civil society but eschewing 'vertical' statist politics. This framework reflects for many the forum's roots in 'the effective disintegration of the old left and its replacement by a more inchoate, plural and diverse set of progressive actors', a development which has 'found its organizational expression in the WSF and its associated "politics of the open space"' (Vanaik 2004, 60).

On the other side of the debate, many lament the lack of political proposals and coordinated political action, arguing that autonomy and consensus decision making is a fraught route to social change. Instead they promote a political movement ready to embrace 'practical' political power. To do otherwise, they argue, is to invite failure. One commentator expressed a widespread view:

> The result of this exclusion of parties and state, if pushed through, would severely limit the formulation of any alternatives to neo-liberalism, confining such aspirations to a local or sectoral context … while giving up any attempt to build an alternative hegemony, or any global proposals to counter and defeat world capitalism's current neo-liberal project (Sader 2002, 92).

One of the architects of the open space model, Chico Whitaker articulates well the distinction between space and movement. He claims a movement 'congregates people … to accomplish, collectively, certain objectives' through an 'organizational structure [that] is necessarily pyramidical'; a space on the other hand, 'has no leaders' and is 'basically horizontal' (in Callinicos 2004, 107). In short, Whitaker cautions against a return to vanguardism – the danger inherent in transforming the forum into a fully fledged political movement – even as many would like to see it transformed into just that. As Wallerstein points out, whether the forum can assume the form of a political movement and 'still maintain the level of unity and absence of an overall (inevitably hierarchical) structure is the big question of the next decade' (2002, 37). Answers to this question will inevitably shape the forum's future.

The debate over autonomous space resonates the politics of Hakim Bey's 'temporary autonomous zone'. As we saw in Chapter 1, the spatial politics of the TAZ inscribe temporary spaces outside state control as revolutionary moments that dissolve domination and enable autonomy. Whitaker's acephalous, horizontal forum space is just such a moment. He utilizes a Beyian-like image of a town square to underscore it: a forum space is 'like a square without owner – if the square has an owner other than the collectivity, it fails to be a square, becoming a private territory' (Whitaker in Callinicos 2004, 107). The space model also echoes the anarchical refrain of fusing means and ends, where the process (autonomy and radical democracy) should match the desired outcome (autonomy and radical democracy). Other commentators offer important insights. Tormey, for example, argues that the WSF encapsulates a horizontal 'utopian space', one that 'opposes vertical politics' and renders possible a 'local, particularistic, individual' politics (2005, 402). Importantly, this is a position that 'represents the dissolution of ideological politics' since it 'accepts, indeed celebrates, the desirability of developing spaces in which we can encounter others on terms that are not mediated' – a 'dialogic' politics that arises from the '*critique* of ideology' itself (2005, 402). This is an important point – and one that goes to the heart of our conceptualization of a post-ideological anarchist influence on radical politics. Tormey points out that if the WSF becomes 'majoritarian' it would cease to be a 'non-denumerable space', a space of minorities, becoming instead:

> ... a vehicle for the expression of a distinct political project that some identify with and others do not. It becomes a party-in-the-making, with a membership, division of labour, leaders and led, manifestos and programmes, exclusions and micro-fascisms ('and at this plenary another of our leaders will be speaking on the subject of ...') ... 'Moving forward', 'answers', 'ideology' and 'pretence' on one side. 'Innovation', 'creativity', 'celebrating spontaneity', on the other (2005, 406).

He captures well the ideological (or more precisely, non-ideological) contest for the forum's heart, a contest that resonates the penetration of a post-ideological anarchist termperament in the forum process and in the core of its radical politics. But where to from here for the WSF is yet to be unveiled.

Conclusion

In introducing a recent collection on the future of the AGM, Hannah Dee (2004, 7) incorporates most of our identified AGM characteristics in her charge that:

> We have succeeded in launching a serious and sustained challenge to a system which puts profit before people and war before peace. We have opposed this agenda in every corner of the globe, at every opportunity presented. World leaders can no longer meet without mass protests organised on an unprecedented scale. We have sent a signal to them and ourselves that we will not be turned back. We will not be dissuaded. We do not recognise their authority, where that authority brings so much suffering and waste. We have set a new agenda and we aim to fulfil it.

Our overview of contemporary oppositional politics has gone even further, however. We have also traced in both the AGM and the WSF a radical politics inspired by anarchist principles and animated by its organizational insights. There are of course diehard anarchists in both movements. But, more importantly, both movements are inspired by a post-ideological anarchism that shun centralized and hierarchical organizational structures, defend a decidedly acephalous character and champion a politics and practice of diversity, plurality and autonomy. While not universally embraced, this anarchist impulse has at the very least stimulated a vigorous debate about the underpinning principles and practices of radical social change.

We have referred throughout this chapter to the role of new technologies in enabling the AGM's network form. We now turn to exploring in more detail the links between the new technologies and the character of contemporary dissent.

3
Technologies of Dissent

It is no surprise that contemporary communication technologies have altered the social fabric in the early 21st century. The surprise would be if they had not. Technological developments have driven major economic, social and political transformations throughout history. Thus the contention that 21st century dissent is also shaped by prevailing communication technologies is a very reasonable one. The internet is central to these technologies. Some contend that the internet is the very 'fabric of our lives'; if we accept information technology as 'the present-day equivalent of electricity in the industrial era', then the internet could now 'be likened to both the electrical grid and the electric engine because of its ability to distribute the power of information throughout the entire realm of human activity' (Castells 2001, 1). But technology is not neutral, nor does it arise in a social vacuum. It is shaped by social forces that determine its form, utility and distribution, and it reflects and reinforces existing power relations. Yet with appropriation by the powerful can come degrees of counter-appropriation. As we saw in Chapter 2, some of the technologies of globalization have been counter-appropriated by globalization's opponents, helping to organize resistance against it. Paradoxically, the shape of the anti-globalization movement (AGM) is in part both a reaction against globalization and a product of the tools made available by it.

The economic heart of globalization is the mobility of financial capital, a mobility enabled by new technologies. Computer-mediated technologies helped provide a solution to the capitalist crisis of a quarter century ago by enabling its more flexible global form. But anti-globalization protest is also enabled by the capacity to spread its opposition cyber-globally through these very same technologies. If global corporate capital is empowered by new technologies then so too is its anti-corporate resistance – even if capacity is far from equal. The new technology enables not only extended message reach but a new style of protest – a networked ecology of dissent that is acephalous, decentralized and particularized. It is here that we find its anarchical temperament. The technology context of contemporary dissent is both relevant to this book's themes – and a very interesting one. An array of computer mediated communication technologies underpins the movement's capacity to practice its dissent. To capture this, the chapter is divided into two main and quite different sections. The first explores synergies between anarchist principles and the cultural environment of early technological developments, particularly of the internet. The internet is often described as 'anarchical' – albeit in a very general, and often disparaging, way. But the internet's early development reveals an embeddedness in a quasi-anarchical 'gift' culture driven by free and open access to an informational commons. While its eventual commercialization ended this momentum, elements of the AGM are determined to retain this ethos. The second section explores more directly the utilization of the new communication technologies in the shaping, driving and mirroring of anti-globalization politics.

The issue of technology, and the utilization of its electronic tools, is a contentious one – both inside the anti-globalization movement and within anarchism itself. We saw in Chapter 1 how the anarcho-primitivist strand of new anarchism is passionately against all technology. These sceptics discern in technology only its 'dark' side. While not necessarily embracing the primitivist line, many radical ecologists have long conceptualized environmental problems in these terms. Not simply against technology, they oppose an instrumentalist and rationalist bias that disembeds the technology from its social and ecological consequences. They reproach the appropriation of technology's profits by a few, and the displacement of the more negative externalities to the many, including the earth itself. Ontological anarchist Hakim Bey observes these contradictory attitudes to technology and the internet within radical politics, noting its two opposing strands:

(1) What we might call the *Fifth Estate*/Neo-Paleolithic, Post-Situ Ultra Green position, which construes itself as a luddite argument against mediation and against the Net; and (2) the Cyberpunk utopianists, futuro-libertarians, Reality Hackers and their allies who see the Net as a step forward in evolution, and who assume that any possible ill effects of mediation can be overcome – at least, once we've liberated the means of production (2003, 108).

While unashamedly using it, many in the AGM are equally circumspect about technology's power relations and its culpability – both as a toolkit and a mindset – in driving social and environmental ruin. The AGM acknowledges that the technological 'cornucopia' is utilized by many – not all of them AGM sympathisers – and technology's uses and developments reflect a complex and dynamic interplay of power politics.

A technologically-charged oppositional politics does not signal a replacement of 'old style' face-to-face interaction and organization, however. Nor does it set out to. It recognizes that both forms are critical to shaping and cohering dissent. There is clearly a spectrum of views on the ideological underpinnings of technology, and how these underpinnings should condition its use. But it is equally clear that communication technologies have significantly conditioned the scope and tenor of 21st century dissent – a development this chapter traces.

Politics, technology and the internet

As developments embedded in and reflecting a society's existing power relations, technological developments have always been political. Throughout history, technologies that serve the interests of the powerful have been selected and advanced. More recently, the interests of capital were served by communication technologies' potential to 'rescue' capitalism from the crises it confronted in the last few decades of the 20th century. The economic crisis of the 1960s and 1970s pinned capitalism's survival on a flexible 'regime of accumulation' (Harvey 1989, 124), one enabled by new computer-mediated technologies. These technologies enabled the emergence of global networks and an 'informational mode of development' that invests increased power in those able to converge 'information technologies and information-processing activities into an articulated techno-organizational system' (Castells 1989, 19). Those best able to take advantage of this new environment were multinational corporations who found that the technology enabled a powerful global interconnectivity that considerably expanded their control and reach.

Echoing a now well established view, Castells (2001, 1) claims the internet as 'the technological basis for the organizational form of the Information Age: the network'. As the organizational basis of communication technology's new age, the network both characterizes and drives a globalized economy and culture. The key features of these 'interconnected nodes' or networks – flexibility, adaptability, horizontalism, decentralization and complex coordination – have significantly reshaped organizational form and function, and help drive globalization (Castells 2001, 2). As a global communication network, the internet is key. Its capacity for speedy interconnectivity and information dissemination has transformed economics, politics and society in most parts of the world. While there may be limited penetration of computer technology in the south, the fact remains that its widespread use in the north impacts significantly even on those societies denied access to it. Computer technology now underpins the practice of globalization. As a communication tool, the internet compresses time and space to allow instant communication to millions of users across the globe. It generates a plethora of virtual global communities that connect, communicate and coordinate in historically unprecedented ways. Since all technologies are both socially produced and socially transformative, the networking function of this latest technology has helped produce new patterns of social interaction and a new economy.

The internet has approximately one billion users worldwide, a figure projected to increase even more rapidly in the near future, driven by the new generation of digitized technologies. This reach is still skewed towards the north rather than the south. According to Internet World Stats (2004), December 2004 figures indicate that internet penetration throughout the world is significant and rapidly on the increase. Internet penetration as a percentage of population in North America is 68.3 per cent, in Europe 31.6 per cent, and Oceania/Australia 48.5 per cent. However, in the developing regions of the world, the figures are significantly reduced. In Africa usage is only 1.4 per cent, in Asia 7.1 per cent (even as the vast population contained in this category represents 31.7 per cent of world users) and Latin America/Caribbean 10.3 per cent. Thus with a combined total of approximately 17 per cent of world population, Europe, North America and Oceania/Australia consume 57.6 per cent of global internet use. As a communication 'revolution' the internet approximates the impact of the printing press. Internet usage, and associated computer mediated communication, has transformed the process of not just economics, but much social interaction.

Whether viewed as utopia or dystopia, these technologies help characterize the lived experience of many in the 21ˢᵗ century, more particularly in the north. For many of those in the north with access to it, the internet may operate non-hierarchically. Paradoxically, however, it can reinforce hierarchy in the south since internet resources are necessarily translated to the masses by the small elite that has access to them. But the view that considerable choice remains regarding the use or non-use of these technologies is misleading. Rather than simply representing an additional extra to the already extensive range of communication 'toys' available to well-resourced consumers, the fact is that exclusion from computer-mediated technologies 'is one of the most damaging forms of exclusion in our economy and our culture' (Castells 2001, 3). Disengagement from online networks invites a spectrum of economic, political, social and cultural consequences.

The libertarian internet

The internet developed in a surprisingly libertarian, anarchic culture – a uniquely cerebral environment that valued freedom, creativity and collaboration. In her absorbing history of the internet, Abbate notes that its history is an interesting and unexpected one: it is not 'a story of a few heroic inventors' but rather one of 'collaboration and conflict among a remarkable variety of players' (1999, 2–3). This knowledge community was motivated by ideas and ingenuity rather than commerce and profitability. It was 'a culture which had no professional secrets, in which co-operative effort was the order of the day and in which the only judgment worth bothering about was that of one's peers' (Naughton 2000, 196). The internet's development and evolution was community rather than individually focused. Key individuals do stand out, but often for their disdain rather than embrace of individual glory. A key contributor to the internet's operational context, Paul Baran, highlights the technology's 'ecology':

> The process of technological development is like building a cathedral. Over the course of several hundred years new people come along and each lays down a block on top of the old foundation, each saying 'I built a cathedral' ... Then along comes an historian who asks, 'Well, who built the cathedral?' ... If you are not careful you can con yourself into believing that you did the most important part. But the reality is that each contribution has to follow onto previous work. Everything is tied to everything else (Baran in Naughton 2000, 77).

In this way, the internet's development is underpinned by a 'gift economy' driven by 'geek' and 'hacker' cultures motivated by intrinsic, rather than instrumental, appreciation of the technology and its experts. As would be expected, the dizzying economic prospects of the new technology soon spawned vigorous entrepreneurial pursuits that mined its immense capacity for wealth generation. Despite this, the technology's intrinsic features permitted parallel non-commercial uses.

The internet story began in the 1960s in the US Defence Department's Advanced Research Project Agency (ARPA) unit. The context was the cold war and the promotion of US strategic interests in the game of brinkmanship that characterized the conflict. While the internet was not originally conceived as a communication tool *per se*, the objective was intelligence enhancement through the development of inter-computer information sharing. In collaboration with a range of other technological experts, ARPA launched ARPANET – a single network connecting several dozen nodes. Utilizing an independently developed and revolutionary new technology called packed switching, ARPANET was able to share information with a number of research units through an interconnected computer network – the first iteration of the internet we know today. The military did not seek to advance the commercial capabilities of the new technology, focusing instead on its military capabilities. But while the institutional arrangements within the military drove the technology's development, at the same time the technological wizards it employed constituted their own community, one that incorporated quite separate and distinct values. The internet was thus born 'at the unlikely intersection of big science, military research and libertarian culture' (Castells 2001, 17). Since 'the kinds of social dynamics that we associate with the use of networks also [come] into play during their creation' (Abbate 1999, 4) these social dynamics produced, in the case of the internet, a libertarian ethos that guided its evolution – for a time at least.

The military itself was steeped in a cultural environment conducive to research and technological innovation. This was assisted to a considerable degree by the extensive financial resources available to it. Innovation was not to be limited by money. Coupled with the research tradition of the participating universities, this knowledge environment created the underpinnings of the emergent 'hacker' culture. The military eventually separated itself from the academic community and created its own server. But it left behind an independent or 'civilian internet', albeit one that eventually took a commercial route (Abbate 1999, 142). The military, ARPA, various universities and think tanks, and different scientific communities made critical contributions to the architecture of the internet. However, many grassroots contributors developed the underpinning software which they then launched – free of charge – into cyberspace. This lay the groundwork for the 'gift economy' that characterized a good part of the internet story. For Castells this represented a 'communitarian approach to technology' where 'the meritocratic gentry met the utopian counterculture in the invention of the Internet, and in the preservation of the spirit of freedom that is at its source' (2001, 33). The ascribing of a communitarian and libertarian ethos to the internet is an important claim, one that goes to the heart of its anarchical impulses.

The internet's libertarian ethos was expressed in a number of ways and manifested in its several intersecting layers. Four primary layers are commonly identified: the techno-meritocratic, hacker, virtual communitarian and entrepreneurial cultures (Castells 2001). While these layers were in many ways hierarchically ordered, it is in their interrelationship – particularly their openness and free exchange of ideas and software – that helped shape the internet culture. The techno-meritocratic culture of scientific, technological and academic experts originally tasked with security objectives and underpinned by a modernist respect for technological progress, occupied a dominant position in the creation of the internet, for a while in any case. In their respect for peer review and dialogue – classic academic research values – and their commitment to building a vibrant knowledge community, these 'techno-elites' nonetheless promoted values consistent with an ethos of open and shared knowledge exchange. Regardless of direct intent, this ethos contributed to the 'democratic' evolution that the internet eventually underwent. The internet culture is thus founded on the academic tradition of pursing knowledge for its own sake, building reputation through peer feedback and review, and the sharing of knowledge and individual research findings with one's community of scholars – values eventually transferred and perfected by the ensuing 'hacker culture' (see Castells 2001, 40).

Labelling the evolving hacker culture the 'great unwashed' or the 'homebrewers', Naughton (2000) points out how these computer whiz 'outsiders' eventually injected the internet with its distinctive anarchic qualities. These computing outsiders 'gazed enviously at the ARPANET and Usenet much as poor kids in black neighbourhoods might peer over the wall at rich kids enjoying the facilities of an expensive tennis training complex' (Naughton 2000, 185). A set of values that respected the intrinsic pursuit and sharing of knowledge underpinned the originating ARPANET community. But it was the 'outsiders' who fully appreciated the internet's 'anarchy' and absence of central control. In this way, while the 'original ARPANET community was a disciplined, orderly, relatively homogeneous, anal-retentive group … their descendants are brash and undisciplined and … beyond anyone's control' (Naughton 2000, 192).

It is thus in the third layer – the hacker culture – that the internet's libertarian and democratic ethos is significantly honed. Hackers now have bad press and their identity and values are sullied. The term hacker originally derived from a 'hack': someone who used technology creatively and innovatively (Jordan 2002, 120). But this original definition has given way to the hacker association with criminal and sociopathic activity rather than technological creativity. Hackers are thus commonly confused with 'crackers': those who seek to create computing chaos or undermine systems either for personal gain or out of a 'geekish' pursuit of 'power-games'. Drawing from the values of the techno-elites, hackers are instead those technological devotees who are motivated by the intrinsic pursuit, creation and sharing of techno-logical knowledge. These creative pursuits are underpinned by an ethos of freedom that manifests as a gift culture:

> Freedom combines with cooperation through the practice of the gift culture, eventually leading to the gift economy. A hacker will post his or her contribution to software development on the Net in the expectation of reciprocity. The gift culture in the hacker world is specific *vis-à-vis* other gift cultures. Prestige, reputation, and social esteem are linked to the relevance of the gift to the community … In addition, there is also gratification involved in the object of the gift. It not only has exchange value, but also use value. The recogni-tion comes not only from giving but from producing a valuable object (innovative software) (Castells 2001, 47).

This is well exemplified in the development of a free UNIX type operat-ing system by a young Finnish undergraduate student, Linus Torvalds, in 1991. This free unit effectively by-passed the 1984 copyrighted, and pricey, UNIX system then owned by AT & T (see Wayner 2000). Interestingly, legislation prior to 1984, when AT & T officially acquired ownership of UNIX, barred the company from commercial activities, mandating that it make its operating systems freely available to univer-sities and research institutions (Naughton 2000, 193–206). UNIX's con-sequent commodification by AT & T thus enraged many in the hacker culture. As Naughton (2000, 198) explains:

> Suddenly the operating system which generations of computer science students and hackers had taken for granted became a product just like any other. It was also priced accordingly – and operating systems never came cheap. And the freedom to tamper

with the system's source code – and redistribute the results – which graduate students had taken as their birthright was suddenly taken away. It was a disaster, an outrage, a scandal.

Hackers such as Stallman and Torvald responded to this commercialization 'outrage' by undermining it.

Launching the *Free Software Foundation* in 1985, Stallman created a 'copyleft' licensing system that effectively undermined this increasing commodification trend. The Open Source movement (see Raymond 2000) also went some way towards retaining openly available software, but its more market-friendly approach invited caution from committed libertarians such as Stallman and Torvald. The latter recognized that while maximum freedom lay in the launch of uncopyrighted programs that users were free to use and alter, there needed to be a way of preventing unscrupulous users from copyrighting and hence commodifying their changes (Naughton 2000, 197). Stallman's innovation was to provide the public with free software under a licensing agreement that permitted them to use and alter the program as they saw fit, so long as this freedom was always passed on to other users (Naughton 2000, 197; Wayner 2000). To distinguish it from copyright, Stallman called his system copy*left*. According to the Free Software Foundation (n.d.), the Foundation is:

> dedicated to promoting computer users' rights to use, study, copy, modify, and redistribute computer programs. The FSF promotes the development and use of Free Software, particularly the GNU operating system, used widely in its GNU/Linux variant. The FSF also helps to spread awareness of the ethical and political issues surrounding freedom in the use of software.

Once the licensing conundrum had been resolved, Stallman and then Torvald had to find software 'which would be worth liberating'. They did this by creating a clone of the AT & T owned UNIX operating system called GNU (Naughton 2000, 197–200). They were not able to produce an operational key element, however, until Torvald launched his UNIX type alternative – Linux – as copylefted software in the public arena (Wayner 2000). By 1992 this alternative system was fully operational, so that by 1998 there were close to 10 million users and 10,000 programmers taking advantage of free access to this software (Naughton 2000, 202). When Naughton asks what enabled this 'extraordinary development' he answers it by identifying a combination of three elements: the copyleft licensing system, the internet itself, and 'the distinctive mindset of those who work on the Linux kernel' (2000, 203).

Demonstrating its libertarian impulses, the Open Source software movement promoted by Eric Raymond conceptualized the creation of software as a open and non-linear 'bazaar' which he contrasted to a more hierarchical and linear cathedral or closed source model (Raymond 2000). Raymond's software model was not necessarily anti-corporate, however, and he was not adverse to commercialization altogether – in the way that the Free Software Movement may have been (see Wayner 2000; Truscello 2003). This helps explain the tensions between the linked Free Software and Open Source movements. Some nonetheless find in Raymond's bazaar model of technology considerable resonance with facets of poststructural anarchism, particularly with Bey's temporary autonomous zone (TAZ) (Truscello 2003). Truscello contends that since the cathedral, or Net, represents a hierarchical information structure and the bazaar a non-hierarchic Web, the bazaar becomes an excellent TAZ space, a 'place where the databanks of postindustrial capitalism can no longer trace your purchase, where technologies of surveillance are not situated, where autonomy is the essence of human communication' (2003).

Rather than simply consumers of a product, these diverse users, and programmers, play an active role in defining the technology's very features (Abbate 1999, 6). But as with the technology itself, these internet cultures are also undergoing their own transformation, as noted in the diverse world of hacking. Political hacking, for example, derives from the hacker culture but consciously targets specific political ends. 'Hacktivism' goes further. Unlike the more straightforward goals of political hacking, hacktivism is motivated by an underpinning libertarian ethos that posits the 'free flows of information, securely accessible to all' as 'the highest principle' (Jordan 2002, 121). These highly politicized hacktivists – among them AGM dissenters – create and utilize new forms of cyber protest and new means of collecting, dispersing and politicizing information.

The next two layers – the virtual communitarians and the entrepreneurs completes the cultural topography of the internet. In exploring the virtual communitarians, Castells (2001, 52) warns that while the values of the technology's producers was central, this should not overlook the contributions of the technology's users in determining its ethos and its shape (see Abbate 1999). These users were not so much interested in the technology's freedom as the communication opportunities it provided. These opportunities included chat rooms, mailing lists, newsgroups, conferencing and multi-user games. For this 'communication community', open communication represented its own value. While many users of the enhanced communication mediums were also skilled programmers, many were not. Yet these users, including AGM dissenters, helped shape the internet's identity. In short, 'while the hacker culture provided the technological foundations of the internet, the communitarian culture shaped its social forms, processes, and uses' (Castells 2001, 53).

The growth of personal computers, the increasing utilization of local area networks, the development of protocol software and, critically, the emergence of the World Wide Web as a more attractive multimedia interface in the early 1990s, launched the internet explosion. One of the Web's creators – Tim Berners-Lee – drew from the 1974 *Computer Lib* hacker 'manifesto'. This manifesto sought to democratize computer use so that ordinary people as well as the 'computer priesthood' could utilize the decentralized participation that it enabled, as well as encouraging users to contribute new content and new software to its evolution (Abbate 1999, 214–20). These developments helped proliferate the open communication culture. The users as well as the producers of the technology contributed to its identity and a plethora of virtual communities proliferated. However, unlike the hacker culture which shared a set of values these communities are immensely varied – with many of them promoting the antithesis of libertarian values. Yet they share a commitment to the value and practice of free, open and horizontal communication – the virtual practice of global free speech. Online communication networks that enable isolated individuals to embrace and be embraced by a virtual community of the like-minded, makes such virtual free speech possible. The social practices inscribed by these users influenced the internet's future shape, helping promote communication tools such as electronic mail and widening its access considerably.

The entrepreneurial culture was, surprisingly, a late developer, although its eventual ascension was rapid. While communication technology helped drive the economic and social character of globalization, at the beginning its potential was not widely embraced. Many of the entrepreneurs who tapped the broader potential of the new technologies are now fabulously wealthy. The floating of Netscape Communications Incorporated on Wall Street in 1995 exemplifies the enormous financial and entrepreneurial phenomenon that was the Net. Describing this phenomenon, the *Wall Street Journal* observed that 'while it had taken General Dynamics forty-three years to become a corporation worth $2.7 billion, Netscape Communications had achieved the same thing in about a minute' (in Naughton 2000, 252). The rest, as they say, is history.

While anarchical impulses may have resonated through the technology's development, it was rapidly corralled for commercial purposes and in the broad service of globalization. It was an entrepreneurial capital assisted by technological reach that helped drive the new economy. Ideas were now as keenly fought over as tradable commodities were in the past. But there remained considerable resistance to the commercial enclosures of the cyberspace commons, and a determination to retain the freedom that sat at the heart of these technologies. Oppositional politics utilized the new technologies to launch their own 'ideas' and inscribe their own forms of dissent. It is to these tools of dissent that we turn next.

The tools of dissent

The new technologies can, on a practical level, make limited resources go a long way, especially for financially stretched protestors. The internet's logistical capacities in helping coordinate large transnational anti-globalization events, are impressive (see della Porta & Mosca 2005). Most organizations have their own internet site on which they broadcast their own specific interests as well as conducting activities such as online forums, campaigns and petitions. But beyond this, computer mediated technologies make available a considerable technological toolkit from which oppositional politics is crafted (see Starr 2000; Arquilla & Ronfeldt 2001; Castells 2001; Klein 2001; Webster 2001; Haythornthwaite & Wellman 2002; Jordan 2002; Pickerill 2003). On a very basic level, the internet provides opportunities for online publishing. There is a proliferation of sites that collect, collate and publish the works of many oppositional movements. The anarchist movement, among many others, has contributed many information sources to this electronic and vast information source. Many previously limited editions of specific radical texts are made freely available on a number of sites and a sizeable proportion of new anarchist literature and other writings are also freely accessible online. The internet also enables an alternative source of news reporting, focusing on news items that seldom reach the mainstream press, as well as providing clarification of mainstream media 'distortions' that impact on the way oppositional politics is reported. Indymedia, for example, is a very well utilized alternative and 'democratic' media source that both collates and disseminates a diverse range of radical news. But this expanded accessibility applies to many other 'suspect' sources of information, and is a primary characteristic of internet technology.

The internet also provides opportunities for political 'hacktivism'. Some of the more straightforward tools of political hactivists include various forms of 'e-protest' such as the flooding of targeted official websites with emails or 'e-occupation' of the virtual web space by concentrated usage. 'Blogging' aims for a virtual shutdown of the website or at least to slow down its availability to other users. Other sophisticated forms of political hacktivism are more direct in their impact. For example, during preparations for the S11 protest in Melbourne in 2000, hackvitists managed to divert all those who logged onto Nike.com directly to the S11 Alliance site, netting S11 over 90,000 hits in nineteen hours (see Burgmann 2003, 295). In addition, these 'cyber-warriors' were able to penetrate other sites and lodge announcements of the upcoming S11 protest on them. This rendered the S11 website one of most popular global sites in the lead up to the September protest.

The new technologies have also infused 'culture jamming' with a new edge. Through the defacing of both on- and off-line advertisements and billboards, 'subvertising' or 'adbusting' dramatically undermines the advertisements' original message. Combining cleverness, humour and creative astuteness, subvertising activists utilize the same advertising techniques of the originals to subvert the targeted message. Through juxtaposing the 'signs' of a contemporary branded capitalism, these agents undermine both the specific value of a given commodity, as well as the highly commodified culture which gives it form. As Klein contends, the 'most sophisticated culture jams are not stand alone ad parodies but interceptions – counter-messages that hack into a corporation's own method of communication to send a message starkly at odds with the one that was intended' (2001, 311). The strategy of culture jamming is not necessarily new, but it becomes even more effective in an increasingly commodified culture driven by emotionally charged brands and logos. Many AGM agents thus embrace the tools of 'detournement' and 'the spectacle' against a contemporary brand-fuelled capitalism. In doing so, they marry the postmodernist political aesthetic of earlier decades with today's more sophisticated technological environment. The new technologies make today's culture jamming both more resonant and more accessible. Nowadays culture jammers have access to a wide variety of software programs that allow them to 'match colors, fonts and materials precisely', with one 'jammer' claiming that there are 'so many different techniques [today] that make it look like the whole ad was reprinted with its new message, as opposed to somebody coming at it with a spray-paint can' (Klein 2001, 316). 'Adbusters' utilize their significant expertise with digital technologies to wreak sophisticated and creative 'havoc' on many corporate sites. Importantly, level of sophistication borrows 'visual legitimacy from advertising itself', with some subversives attracting – albeit grudging – admiration from their very targets (Klein 2001, 316). We discuss culture jamming further in Chapter 7.

On a broader ideological and philosophical level, the new technologies facilitate an organizational form that resonates the movement's anarchical temperament. This temperament is located primarily in the non-hierarchical and decentralized organizational forms permitted by the new technologies. Moreover, decentralized and non-hierarchical forms of organization are 'selected' or privileged by technologies such as the internet (see Gerlach 2001). These technologies enable networks of dissent to strengthen co-ordination capacity at the same time as respecting the integrity and autonomy of their individual cells. Computer mediated communication (CMC) technologies have, in particular, assisted smaller local groups and individual activists to link themselves flexibly to these cellular network structures (Pickerill 2003). This connection to fluid cellular networks helps compensate for these groups' limited resources, and enables them to participate more fully in the shaping of a formidable ecology of dissent. Klein characterizes this robust and diverse network as 'hubs and spokes' of a broader network ecology where 'rather than a single movement' there is emerging instead 'thousands of movements intricately linked to one another, much as "hotlinks" connect their websites on the internet' (2002, 4). But in contending that this analogy is far from coincidental, she claims that it in fact describes 'the changing nature of political organizing':

> Although many have observed that the changing nature of the recent mass protests would have been impossible without the Internet, what has been overlooked is how the communications technology that facilitates these campaigns is shaping the movement in its own image. Thanks to the Net, mobilsations are able to unfold with sparse bureaucracy and minimal hierarchy; forced consensus and labored manifestos are fading into the background replaced instead by a culture of constant, loosely structured and sometimes compulsive information swapping (2002, 4).

When describing her 'unity of many determinations' Starr (2000, 158) too explains how the new technology facilitates both the movement's diverse and particularistic identities as well as their strategies. It builds alliances premised on diversity and coordinated strength. She connects these diverse alliances – this 'globalization from below' – into a matrix of networked nodes around shared values and objectives. Through a creative and strategic use of the new technologies, these networks help forge the key elements of the movement's ecology. Echoing a widespread sentiment, a prominent eco-feminist claims a direct relationship between the AGM and the internet. Using the metaphor of a web, she contends that it both implies and enables 'a pattern of connections that are complex and flexible', and that circumvents authority and hierarchy through the capacity of any point on the web to 'communicate with the centre' (Starhawk 2003, 171). This social pattern reflects 'a classic spider web' whose 'spokes radiate out from a central point, linked by a spiral of sticky thread' (2003, 171). But this pattern is not just an organizational device – it is also a personal transformation strategy that underpins social change. Since radical change requires changes in ways of thinking as well as ways of being, Starhawk contends that 'we need to learn to see patterns, to think in terms of flows and connections rather than focus on the things they connect' (2003, 260).

While Starhawk speaks of patterns and webs and Klein (2001) of hubs and spokes, others utilize Deleuze and Guattari's distinctly anarchical 'rhizome' analogy to describe these networks of dissent (see Tormey 2004, 160–2). As a permanent underground stem that generates a random profusion of roots, a rhizome's shoots appear to produce individualized and separate plants. In fact these individual plants are connected by the single, but invisible, stem. This parallels the AGM's capacity to incorporate individualized and separate protest 'shoots' – each assuming its own shape and identity – into an interconnected root that opposes neoliberal globalization. However conceived, this network form mirrors the movement's overall ideology. The AGM is an ecology of dissent where autonomous groups come together within the ambit of a larger protest whole of which they are a constituent but independent part. The internet has thus facilitated a form of 'horizontal activism' that incorporates several key features: first, oppositional dissent is driven by a network structure able to bypass bureaucratic and hierarchical organization; second, engagement rather than formal membership defines network participation; and finally, networks generate spontaneous, unpredictable and impermanent formations and alliances (Tormey 2004, 159).

This raises the question of how exactly the anarchical temperament of the new technologies and contemporary oppositional politics intersect. First and foremost, the internet is the tool of 'accelerated pluralism' (Bimber 1998, 2001). While Bimber warns that this accelerated pluralism is used to promote all kinds of ideological projects and is not always a force for 'good', this radical pluralism reflects and drives radical dissent. In accelerating such pluralism, the internet becomes a normative meta-tool of oppositional politics. The internet not only simplifies the incorporation of otherwise marginalized groups into a networked whole, it also enhances both coordination *and* autonomy, generates alternative and accessible sources of information, and creates new forms of direct action (Tormey 2004, 64). Networks such as Indymedia, People's Global Action and Direct Action Network (during the Seattle protest), not only provide information and news about political issues and upcoming events, but help cohere its audience of autonomous groups into a network of shared opposition. While networking of oppositional groups has occurred in the past, the internet has honed this capacity considerably. The Zapatistas are a prime example of how a small, marginalized group, rooted very specifically in their locality, has utilized the internet to propel themselves onto the world's oppositional stage. The Zapatistas' Subcommandante Marcos used the web and the internet very strategically to disseminate Zapatista 'communiques' across the globe. There are now tens of thousands of Zapatista-related websites globally and these communiqués have been translated into over a dozen languages. We discuss the Zapatistas in Chapter 5.

Accelerated pluralism is also enabled by very straightforward tools such as email. Email is a now a preferred mode of communication for many in the developed world, both for business and personal interaction. As an organizational and communication tool, email also has very distinct advantages for activists:

> ... one message can simultaneously be posted to a large number of people; draft text can be shared, discussed and jointly approved, providing a rapid and collaborative way of developing press releases, flyers and reports; emails can be amended or otherwise added to, and then forwarded on to still more people ... And over more dispersed networks, messages can be sent to countless recipients across the globe in 'instantaneous time' (Horton 2004, 739).

At the same time, the immense organizational benefits of mobile phone technology should not be overlooked. Indeed, mobile phones have been pivotal to organizing the large numbers of people who take part in anti-globalization protest events – both before and during the protests. They are critical, for example, in organizing and communicating the last minute organizational locations of Reclaim the Streets events and other street parties. The scale and organizational precision of such large anti-globalization events as Seattle, Genoa or Melbourne would not have been possible without this 'simple' technology.

As the movement's main organizational pivot, the affinity group is well accommodated by the new communication technologies. An organizational form closely linked to anarchism, the affinity group's prime was during the Spanish Civil War when anarcho-syndicalists mounted their opposition to the fascist state. Since then affinity groups have been a hallmark of the new social movements and the mass blockades mounted during the late 1960s and 1970s, especially the civil rights, anti-nuclear and anti-Vietnam war campaigns. Viewed as key components of direct action, autonomous affinity cells choose to work independently or in collaboration with other groups. According to the anti-capitalist collection, Notes from Nowhere (2003), the optimum size for affinity groups is between five and fifteen people. This size helps circumvent the development of hierarchies or centralized oligarchies and allows each member to directly contribute to the group's decision-making processes. While drawing from a long historical tradition, the new technologies have strengthened facets of the affinity group form. These groups move in and out of coordinated protest activity depending on the aspirations of their individual units. They may coalesce for a major protest action, coordinating and collaborating for some time, only to disperse again as they return to their original locations and particularized goals. Although face to face meetings and interaction form the critical encounters, computer mediated technologies help inform, cohere and connect these groups.

While in practice affinity groups were usually organized into clusters for specific actions such as Seattle and Genoa, with 'spokescouncils' formed from groups representatives, these organizational spokes retained considerable autonomy and flexibility. Affinity groups primarily 'belong' to the individuals constituting them and exist outside any permanent, centralizing authority. For many, the bringing together of creative people 'to work and play collectively' is considered one of the movement's 'most powerful acts of resistance' (Notes from Nowhere 2003, 88). In this way, affinity groups form the central hub of global protests and help define the multiplicity of dissent forms that characterize these events. Importantly, the new communication technologies have enabled the speedy coordination of this plethora of activities. This efficient dissent network meets both political and strategic goals. Politically, the affinity group structure underpins the movement's diverse and autonomous anarchical form, corroborating means and ends. Strategically, these groups' impermanent, cellular and acephalous features contrast with mainstream forms of organization, making it difficult for security forces to target and control them. Where the corporate culture champions concentration, the movement favours fragmentation; instead of globalization, the movement practices localization; and against the consolidation of power, the movement pits its radical dispersal (Burgmann 2003, 298).

The new technologies aid the fragmented, decentralized and polycentric organizational form favoured by the many autonomous groups that constitute the AGM. As della Porta and Mosca (2005, 185) conclude:

> [T]he Internet has had, has and will continue to have a meaningful effect on collective action. The Internet empowers a series of fundamental functions of social movement organisations: it modifies their movements' organisational structure (more and more networked, flexible and polycentric) and makes organising demonstrations easier; it increases the possibilities for a direct intervention in politics through different forms of cyberprotest, it influences identity processes and helps to spread alternative information.

Through these enhanced functions, the AGM has been able to launch itself as a successful oppositional movement. Whether these same functions and strategies can retain movement participants is another matter altogether. Caution against over-valorization of the movement's cyberprotest is thus required.

The limits of technological dissent

The movement's decentralized, cellular form may be excellent for attracting participants, but not always for holding them. A resounding objection to leaders or hierarchies can also result in a loss of expertise and focus. In advocating a tactical balance between these qualities, observers urge against a network activism that 'makes a fetish of communication at the expense of action' and that prevents participants from working in 'a coordinated way and in terms of an agreed strategy *for* action' (Tormey 2004, 167). The new technology is undoubtedly effective at connecting many people together across the globe and in coordinating mass global actions. But this does not mean it is automatically good at taking the protest further – 'to the next stage' of changing the political landscape as these protestors want. The equally technologically-savvy opponents of the AGM can quite easily infiltrate activist sites and their organizations. It is worth citing Klein (2002, 9) at length here:

> There is no question that the communications culture that reigns on the Net is better at speed and volume than it is at synthesis. It is capable of getting tens of thousands of people to meet on the same street corner, placards in hand, but it is far less adept at helping those same people to agree on what they are really asking for before they get to the barricades – or after they leave. Perhaps that's why a certain repetitive quality has set in at these large demonstrations ... The Net made them possible, but it's not proving particularly helpful in taking them to a new stage ... Now the police have subscribed to all the e-mails lists and have used the supposed threat posed by anarchists ... to buy up all manner of new toys, from surveillance equipment to water canons.

In his *Postmodern Anarchism* (2003), Lewis Call utilizes a concept linked to electronic gift cultures when talks of 'radical gift-giving' – or of the giving of gifts without expectations of return. Call finds elements of gift-giving in computer discussion boards, interpreting them as 'gifts of advice'; while free and open source software, including music, images and text, are interpreted as gifts of information. Bey too uncovers some interesting links. In 1991 he originally distinguished between the internet and the web (or 'counter-Net'), finding within the internet a more horizontal and less hierarchic inner network he called the web. Like Starhawk (2003), he found the image of an organic and ecological web particularly appealing. More importantly, the web offered significant implementation opportunities for the TAZ. Through its capacity for 'carrying information from one TAZ' and in providing 'the epics, songs, genealogies and legends of the tribe', the web defended the TAZ, 'rendering it "invisible" or giving it teeth, as the situation might demand' (2003, 107). A decade or so later, and Bey is more circumspect on the promises of the new technologies. In the preface to the 2003 edition of *TAZ*, he regrets his former enthusiastic embrace of the

counter-Net and web. He argues that by 1995, the internet 'had suc-
ceeded in burying the "anarchic potential" of the Net' under the
onslaught of a 'triumphalist, evanescent, aesthetically bankrupt, monocultural [and] violent' global capitalism (2003, xi). While many fellow
dissenters would agree, they would not perhaps go quite so far as to
dismiss technological potential altogether.

In short, the technology's anarchical ethos did not prevent its
commodification and commercialization by an eventually dominant
entrepreneurial culture. Free access is a two edged sword. As Sassen
pointed out several years ago, we are at 'a particular historical moment
in the history of electronic space when powerful corporate actors and
high performing networks are strengthening the role of private electronic space and altering the structure of public electronic space'
(1998, 194). Most radicals and activists are now aware of this reality
and the limitations it imposes on their utilization of opportunities
provided by the new technologies. They thus approach them with
caution and circumspection. They note with dismay the increasingly
unsavoury consumption of the technology. Many hackers are motivated by personal gain rather than social justice goals. Many others
seek to create chaos by launching computer viruses in an adrenalin-
fuelled game of technological one-up-man ship. Politically motivated
hacktivism is also utilized by a broad spectrum of players inspired by a
wide range of ideological positions. These range from sites promoting
extreme right to extreme left positions, and everything seemingly in
between. As we saw, the openness of the internet means that movement opponents can also access protest sites, utilizing information
gathered there to undermine the movement. Oppositional politics
now takes place in a considerably reshaped spatial and cyberspatial
environment. But the resilience of many social movements depends in
large part on their strategic accommodation and creative utilization of
these environments.

The contribution of network technologies to the character and politics of oppositional dissent hence remains significant. This significance
reaches to the very identity of social movement politics in the contemporary electronic age. For some, rather than simply representing
'a handy tool to be used because it is there', the internet instead reflects
'the basic features of the kind of social movements emerging in the
Information Age' (Castells 2001, 139). These features include, as we saw,
their acephalous, decentralized, non-hierarchical and cellular nature. In
considering such features, Castells claims these new social movements
as representing a 'pure' and novel form of movement:

[Their] novelty is their networking via the Internet, because it allows the movement to be diverse and coordinated at the same time, to engage in a continuing debate, and yet not be paralysed by it, since each one of its nodes can reconfigure a network of its affinities and objectives, with partial overlappings and multiple connections. The anti-globalisation movement is not simply a network, it is an electronic network, it is an internet-based movement. And because the Internet is its home it cannot be disorganised or captured. It swims like fish in the net (2001, 142).

While segmentary, polycentric, and integrated network structures (SPINs) also described the new social movements of the 1960s and 1970s (Gerlach 2001), the new technologies enhance these capacities considerably.

A tendency towards a zero-sum perspective distorts understanding of the new technological landscape. 'Netwar' (Arquilla & Ronfeldt 2001) may be the new term used to encapsulate and describe the new oppositional environment, but the fact remains that political hacktivism has not necessarily replaced traditional forms of protest. Rather, the available technologies have embellished them. The libertarian protests of the 1960s also challenged hierarchy and centralization, introducing a form of protest carnivalesque that celebrated spontaneity and difference. As with these earlier movements diversity, democracy and participation remain key principles of anti-globalization dissent. The new technologies animate these principles by providing the hard/software to drive them. The affinity group symbolizes this new protest environment well: it is small enough to retain practicable face-to-face communication, yet able to act as an autonomous component of a globally extensive, cyber-networked protest.

The new computer mediated technologies facilitate and expand communication. Once again, they do not – nor do they set out to – replace 'older' forms of face-to-face and other kinds of communication in real time. They do not replace the special bonds engendered by direct, unmediated face-to-face relationships. They do not substitute for the stirring solidarity aroused during a demonstration. Nor do they replace the day to day practical activities of activists in their various local settings. Instead, 'virtual networks operate at their best when they are backed by real social linkages in specifically localized communities' (Diani 2001, 126). In her study of how British environmentalists utilize the internet, Pickerill (2003) highlights the internet's embellishment of traditional activist practices rather than their replacement. In short, the internet is best at building on an already existing pool of social capital, rather than creating it from scratch (Haythornthwaite & Wellman 2002, 29). But it is precisely at the intersection between an active social movement base and available communication technologies that a potent oppositional potential is unleashed – one empowered by a technology that can mirror the movement's ideology and fuel its capacity.

The anti-globalization movement's anarchist impulses are underpinned by this determination to remake democracy through the utilization of non-hierarchical networks. The cyber century contributes to the achievement of these goals. 21ˢᵗ century technologies permit a global coalition of a diverse range of autonomous, independent and only informally affiliated oppositional groups. But of their own accord the new technologies do not herald more democracy or more interconnectedness. Rather the trick is to use them in a way that gives them more democratic form. As Bohman (2004, 140) argues, the internet 'becomes a public sphere only through agents who engage in reflexive and democratic activity'. Rather than simply focusing on its capacities, the internet's democratic utility is determined by *how* it is used. Just because communication technologies have a capacity for democracy does not mean that they will be used for this. The corporate sector promotes the technology's commercial potential while the protest sector its democratic one. The internet 'is a public sphere only if agents make it so'; in short, its character is determined by 'the mediation of agency' rather than 'on the technology' (Bohman 2004, 132, 139).

Conclusion

This chapter has explored the links between oppositional politics, anarchist impulses and the new technologies. It found that internet technologies developed within a libertarian environment underpinned by collaborative, cooperative and democratic values. This electronic 'gift culture' promoted an informational commons operationalized by the free and creative exchange of ideas, infrastructure and software. But it also found that the new technological landscape offers the AGM both opportunities and risks. The internet's anarchical origins reflect the libertarian culture which cultivated and nurtured its communication and informational commons. The corporate culture was relatively late to garner the internet's commercial potential, but has since made up for lost time by rapidly privatizing and enclosing it.

Animated by the non-hierarchical and polycentric capacities of the internet, oppositional groups are nonetheless resolute in resisting its commodification and utilizing it as a tool of accelerated pluralism. Counter-appropriating it for radical purposes, these dissenters are determined to protect the technology's freedom and autonomy. Since 'electronic space is embedded in the larger dynamics of organizing society' (Sassen 1997, 1) and since 'power in the network society are equally located in the architecture of bricks and mortar and the architecture of information' (Truscello 2003, 1) the fight to retain control of electronic space becomes a metaphor for the broader struggle for freedom and autonomy. A post-ideological anarchism helps fuel this struggle.

4
Ecology and Anarchy

Anarchist ideas have long inspired the radical arm of the green movement. Launched alongside the raft of new social movements of the 1960s and 1970s, the green movement has proved one of the most successful and most enduring. Despite public perceptions to the contrary, the green movement, like most movements, is a broad church with a diverse political congregation. As the radical wing of the green movement, radical ecology soon distinguished itself from its reformist arm. Like many social movements of the time, radical ecologists were also inspired by some of the insights of the New Left, and of postmodernism and poststructuralism. They took these insights further, however, charging that the will to power degrades both people and nature. Many condemned not only the androcentric and technocentric values of a modernist industrialism, but also its overarching anthropocentrism. All agreed that domination went beyond class and manifested in a diverse range of repressive practices. This now included the violation of nature. In some form or other, most radical ecologists embraced anarchism's identification of hierarchy as central to the operation of domination. Anarchists too recognized that radical ecology's new insights offered them an opportunity to update their own chronicles of domination.

Radical ecologists form an important component of the anti-globalization movement (AGM). They contribute both to the greening of anarchism and to oppositional politics more broadly. Radical ecology is not a single discourse, however. It incorporates a number of distinct and diverse positions, some of them in vigorous opposition to each other. It may have emerged in the 1960s and 1970s within the broader green movement, but its diverse ideological roots often go back much further. Depending on which position it defends, radical ecology draws from the major ideologies of Marxism and anarchism, as well as an array of philo-

sophical, theological and cultural traditions. Among its eclectic membership we find 'animal liberationists, bioregionalists, eco-feminists, deep
ecologists, social ecologists, eco-Marxists, eco-socialists, eco-anarchists,
ecological Christians, Buddhists, Taoists, pagans, environmental justice
advocates, green economists, critical theorists, postmodernists, and
many others' (Dryzek 2005, 181). Radical ecology includes specific eco-
anarchist schools such as Murray Bookchin's social ecology, which
defends a largely uncompromising position. Others such as Earth First!
embrace the looser influences of anarchism to construct their green radicalism. Still others draw post-ideologically from anarchism those pieces
that help them assemble and strengthen their own specific discourses.
But all radical ecologists distinguish themselves from their more moderate counterparts in the green movement. This explains the generally
accepted distinction between environmentalism (moderate and analytically narrow) and ecologism (radical and analytically extensive) (see
Dobson 2000).

Two features stand out in the contemporary theorization of ecologism. One is the development of environmental ethics and the other,
an infusion of anarchist sensibilities in ecologism. While ecologism
draws from the traditional political spectrum in various ways, it is particularly indebted to anarchism. Many of the values and principles promoted by radical ecology as a whole resonate anarchist impulses. There
is no historical continuity between classical anarchism and ecologism,
despite the 'uncovering' of Kropotkin's green credentials as discussed
in Chapter 6. Rather, there was a mutual accommodation between key
parts of anarchism and ecologism. Anarchism adapted itself well to the
values of the 'ecological paradigm' (Hay 1988) and ecologism drew
from it liberally to strengthen its own discourses. Regardless of 'lineal
descent' or historical continuity, issues close to the political heart of
anarchism have resurfaced through the medium of ecologism. Anarchism had already extended the discourse of domination beyond class
and productivism. In stressing the central role of hierarchy in effecting
domination, modern anarchism was open to including other repressive
practices such as patriarchy and the domination of nature in the
matrix of social control it challenged.

This chapter explores the influence of anarchist ideas in informing
radical ecology. It first traces the contributions of the New Left, critical
theory and postmodernism on shaping the political context in which
the green movement emerged. It then discusses the division of the
movement into mainstream environmentalism and radical ecology,
before focusing in more detail on some of the major radical ecology

schools themselves. It soon becomes clear that greened versions of the Marxist/anarchist debates emerge through the discourses of eco-socialism, eco-Marxism and eco-anarchism. Deep Ecology develops a distinctly novel ecocentric ethic that burrows in at different points of most radical ecology schools. Social ecology, which rejects it altogether, is the exception. While a distinctive school on its own, eco-feminism is informed by the other schools and constructs its various positions accordingly. Bookchin's social ecology is the most prominent of the specific eco-anarchist schools, although recent schools such as anarcho-primitivism also carve out an anthropologically informed social ecology of sorts. We leave Bookchin's social ecology to Chapter 6.

Green politics in the 21ˢᵗ century

The green movement is one of the most successful social movements of the late 20ᵗʰ century. It remains so in the new millennium. Buttel calls it the 'master global social movement' (2003, 99) and Castells (2004, 72) 'the most comprehensive, influential movement of our times'. The green movement not only raised awareness of environmental problems but also placed them firmly on the political agenda. Many environmentalists continue to articulate their concerns as members of a plethora of green civic organizations. Many more are not so connected but demand that their governments respond effectively to accelerating degradation. The past few decades have thus seen an enormous growth of environmental institutions, agencies, green parties, policies and plans across the globe. Most governments today have clearly outlined environmental plans and are aware that many of their citizens cast their votes on the strength of a government's environmental record. Environmental concerns thus occupy a central position in both politics and society. Being green – at least moderately so – is no longer 'fringe' and awareness of environmental problems has generated considerable behavioural change. The absorption of environmental concerns within politics, business and society reflects this success.

For radicals, this mainstreaming of environmental concerns represents the victory of rhetoric over substance. They argue that rather than promoting robust responses to critical environmental risks, state and corporate sectors simply pay a wordy lip service to them. Radical ecologists thus regard the opening decade of this century as one defined by both accelerating environmental risk and accelerating injustice. This helps explain radical ecology's involvement with the anti-globalization movement. While moderate environmentalists may grumble, they are generally heartened to see the global uptake of sustainable development and ecological modernization policies – moderate, technologically focused responses that now dominate the environmental policy agenda (Hajer 1995; Mol & Spaargaren 2000; Curran 2001). Green radicals are not so heartened. They regard the mainstreaming of environmentalism as its death knell. With willing governments usually in tow, they believe that control of the environment agenda has moved increasingly into corporate hands. They highlight a relentless assault on environmentalism by both corporations and governments (see Beder 1997, 2001). For radicals, the corporate assault on environmentalism reflects a powerful counter-strategy by the corporate sector to resume control of an agenda they risked losing. At the same time corporations trumpet their seemingly green credentials in a successful public relations exercise that presents them as good corporate citizens. In short, corporations utilize their considerable resources to propagate a rhetoric of corporate social responsibility that masks the reality of accelerating degradation and scant regard for social and environmental justice.

Radical greens observe this corporate cooptation and the associated seduction of moderate environmentalists with increasing dismay – but a dismay that fuels their determination to resist it. It reinforces their anarchical hunch that any truck with statism invites the absorption of dissent. While the goal of environmental consciousness raising may have been won, it came at a significant price: the loss of control over the very green agenda environmentalists promoted. This environmental appropriation has been largely enacted through the discourses and practices of sustainable development and ecological modernization. Viewed as positive developments for the moderates, and retrograde ones for the radicals, sustainable development is the dominant environmental policy framework today, both nationally and globally.

These developments also reflect, and are reflected in, the internal tussle for control of the green movement. The initial and highly charged success of the green movement in advancing environmental concerns represents a stunning victory for the movement's reformist factions. The global discourse of sustainability is further testimony to this success. While these successes are treated cautiously by radical ecologists, this is by no means the first time internal divisions within the movement have manifested. Internal debates over the movement's identity and its strategy go right back to its origins. Most movements contain a spectrum of political views, even as they unite around common concerns and common goals. The green movement too has always contained its reformist and its more radical arms. Its radical ecology wing was particularly influential during the movement's genesis, especially in the 1970s and 1980s. During this time the green movement was divided into two main camps: mainstream environmentalism and radical ecology. Through the developments of sustainable development and ecological modernization, mainstream environmentalism was able to proclaim its dominance.

This dominance was played out in two interrelated ways. First, as we saw, mainstream environmentalism argued for a reform, rather than transformation, of the existing institutional culture. This resulted in a formal institutionalization of environmental politics that saw significant organizational activities: the evolution of green parties, environmental departments and agencies, and a plethora of environmental policies, regulations and legislation. The focus throughout was a moderate and reformist one. Business increasingly presented themselves as willing partners in this environmental modernization process. Most governments had also turned at least a little green. But the arena of institutionalized politics is a very different one to that of civil society. Even those green groups and parties motivated by broader social justice concerns find that doing successful politics in this arena is a very different proposition from being outside it. The process of policy making in advanced democracies is often a vigorous contest between competing interests. It is particularly difficult to accommodate both principles and power in the real world of pragmatic political bargaining – one of the reasons anarchists avoid it. In this context, the green movement, like many social movements before it, also experienced 'a process of deradicalisation, oligarchisation, institutionalisation and professionalisation' (Van der Heijden 1999, 201). The contest between the 'fundis' (radical idealists) and 'realos' (moderate pragmatists) in the German Greens is testimony to this, especially through the latter's success in pressing the party's pragmatic professionalization and moderate turn.

Second, in the face of such wide ranging institutional successes, the radical ecology component of the green movement was marginalized. Increasingly, the more radical elements were viewed as impractical, uncooperative, utopian and anarchic. Given the considerable energy that was concentrated on formulating and advancing the sustainable development agendas, the green movement's more radical elements were correspondingly sidelined. In the unlikely event that governments would seek their cooperation it would most likely not have been forthcoming in any case. In effect, by the final years of the

20th century, it seemed that the more reformist strain of environmentalism had indeed emerged triumphant over its more radical arm. The considerable strides in sustainable development, the notable growth of many governments' environmental policy portfolios, and the acceleration of a plethora of environmental regulations worldwide, made the continued criticisms of the radical greens seem increasingly petulant.

This did not spell the end of radical green dissent, however. Rather, the radical elements did not so much disappear as go 'underground' to await what they foresaw as the inevitable exposure of the sham that paraded as sustainable development. In any case, radical ecologists such as Earth First!, eco-anarchists and many eco-feminists already kept their distance from the reformists. They believed they would not be so easily seduced by the power the hierarchical state promised them. They would bide their time. Globalization and accelerating environmental degradation provided this opportunity, prompting the re-emergence of an invigorated green radicalism. Radical ecology's gaze was now even more firmly fixed on the link between globalization and inequality and between the trashing of ecology and society (see Curran 2004b; Horton & Patapan 2004). Anarchism rode proudly on the coat tails of this new radicalism, steered and inspired by the anti-globalization movement.

Greening postmodernism

To better understand the state of play of radical ecology in the early 21st century, we need to now go back to its beginnings. Since facets of its early anarchical influences lay here, we briefly trace the impact of postmodernism and poststructuralism on the movement, before examining the separation into its mainstream and radical arms. As we saw in Chapter 1, the New Left and postmodernism after it introduced a wide ranging critique of Marxist materialism along with the modernist values that underpinned it. This helped provide the impetus for the new social movements, including environmentalism. The reflexive political climate also better accommodated the insights of anarchism. Radical ecology found in anarchist thinking many ideas that helped inform their budding ecological discourses. An emerging postmodernism assisted the penetration of these anarchist influences. While this 'assistance' was by no means direct, conscious or even welcome, postmodernism's overarching critique of the dominatory, hierarchical, rationalist, instrumentalist and exclusivist values that underpinned modernism caught the attention of many radicals frustrated by the dominance of modernist values.

Many radical greens accuse modernism's core values of contributing to the desolation of nature. They claim that as a secular movement devoted to the 'liberation' of humanity, or at least some sections of it, from the control of powerful religious elites, modernism instead unleashed other forms of domination. While a self-conscious humanism challenged the perpetual cycle of sin and redemption that dominated the middle ages, and while it relegated man instead of God to the centre of the universe, its impact was nonetheless exclusionary. The modernism that resulted was positivistic, technocratic and rationalistic. Although the goals of human liberty and human happiness may have been commendable ones, critics charged that the price for modernism's ambitions was the invocation of a new language and practice of domination. Of particular concern was modernism's elevation of an instrumental and technocratic rationality as the supreme achievement of human culture. For radical greens, this rationality was instead responsible for man's (literally) domination of nature, now realized through an increasingly sophisticated technological and scientific mastery. There was not only suspicion of the rationalist methodologies employed to first understand nature: Cartesian 'atomism' and the positivist method. There was also a vigorous antipathy for the associated discourses and technologies that drove this mastery.

Some radical ecologists promote these views more so than others. Deep Ecology identifies anthropocentrism as the most problematic Enlightenment value, while eco-feminists highlight patriarchy. Either way, both implicate hierarchy as the driving ethos. This hierarchy replaces God with man but invests him with dominion over women and nature. Radical ecology thus challenges many of the values that have guided western culture since the Enlightenment. As eco-feminists point out, there is an 'excluding quality [to] the supposedly universal "we" of much humanist discourse' – one that is not inclusive of women or, for that matter, of other cultures and peoples, entrenching instead an 'ethnocentricity' of western humanism (Soper 1990, 11). In privileging man to the exclusion of women, 'white' to the exclusion of 'black', and culture to the exclusion of nature, the die was cast for the epistemological and political challenges that would eventually emerge. Radical ecologists found the androcentrism and anthropocentrism at the heart of modernism particularly alarming.

From the radical ecology perspective, postmodernism rebels against the anthropocentric, androcentric and ethnocentric foundations of modernism. Modernism's rationalism and instrumentalism hides a logic of domination whose repression is widespread. Instead of an exclusionary modernism, postmodernism invites 'a theoretical celebration of difference, a resistance to all synthesizing discourse, an assertion of an infinite and multiplying plurality of particulars and specificities' (Soper 1990, 11). Adorno and Horkheimer (1972) argued similarly in their influential *Dialectic of enlightenment*, where they located the tentacles of domination beyond class and into the cultural, the technocratic and the aesthetic. Such thinking resonated radical ecology's suspicion that there was a very strong connection between the domination of nature and the domination of people. If in different ways, both postmodernism and radical ecology attempt to overthrow the constraints of a universalist ethic that has domination at its core.

Postmodernism and poststucturalism have exerted considerable influence on contemporary anarchist thought and radical ecology. We saw in Chapter 1 how some contemporary anarchist schools have clearly embraced some postmodern and poststructuralist insights to fashion their own distinctive analyses. Nonetheless, radical ecology's rejection of some central modernist values does not mean an automatic leap onto the postmodernist bandwagon. There remains considerable suspicion among greens of postmodernist posturing, and these discourses have not been embraced uncritically. Many radicals remain decidedly circumspect, identifying in these discourses an apolitical and non-visionary stance that goes against the impulse of radical social change. Albeit for different reasons, both social ecologist Murray Bookchin and anarcho-primitivist John Zerzan refute the 'trappings' of postmodernism. But in postmodernism's identification of the enlightenment's dominatory core; in poststucturalism's identification of the multiple disciplinary practices that enact this domination; and in the launching of 'the politics of difference', radical ecologists have found many analytical and political tools with which to construct their own discourses.

From mainstream environmentalism to radical ecology

While both environmentalism and ecologism promote conservation values, their proposals are very different. Dobson (2000, 13) defines environmentalism as 'a managerial approach to the environment within the context of present political and economic practices'. As reformists, environmentalists adopt a business as usual approach, albeit a considerably greened one. As we could expect from radicals, their goal is transform society by changing core values. But at environmentalism's outset, much of what it argued for was comparatively radical.

Environmentalism constituted new thinking and new territory, and challenged the pace and process of capitalist economy in unprecedented ways. Many of the environmental issues it raised had never been considered before. While ecologism has a considerable historical pedigree, from 19th century green romanticism to Kropotkin's scientific ecology, the considerably reshaped emergence of these concerns represented a sweeping development that penetrated public consciousness. In also generating considerable alarm, it prompted the movement's initial 'eco-doom' or 'survivalist' phase (Dryzek 2005, 27–50).

In the early 1960s Rachel Carson's groundbreaking Silent Spring (1962) spearheaded the new environmental consciousness. Concentrating specifically on the effects of chemicals (especially DDT) on both human and non-human life, and on the complex interconnections between natural and social ecology, Carson's observations drew the public's attention to the previously overlooked issue of environmental degradation. Driving this new awareness was the realization that human health and human interests were at stake. Environmental protection was hence necessary to protect humanity against the toxic ravages that unregulated development imposed. Allied to the human interest agenda was the consequent economic impacts of resource depletion and degradation. Both people's and the economy's health was at stake. Issues of pollution, population explosion, finiteness of resources, nuclear power plants, and toxic waste were quickly catapulted to the forefront of environmental awareness. The publication during this period of *The Limits to Growth* (Meadows *et al.* 1972) and *Blueprint for Survival* (Goldsmith *et al.* 1972) both heralded and symbolized the emerging survivalist theme. The sub-tile of *The Limits to Growth: A Project on the Predicament of Mankind* exemplified well the portent of doom that menaced humankind.

This phase of environmentalism concentrated on issues of resource depletion and population pressures and propagated a number of authoritarian responses to perceived ecological crisis. Books such as Ehrlich's *The Population Bomb* (1968) and Commoner's *The Closing Circle* (1971) lay the groundwork for the rapid growth of 'survivalist' literature. Ehrlich and Commoner did not proffer the authoritarian responses of other influential works such as Hardin's 'The Tragedy of the Commons' (1968) and Ophuls' *Ecology and the Politics of Scarcity* (1977). Hardin best exemplifies the authoritarian, survivalist tradition. He defended a Malthusian argument: resources are finite and limited and in a grossly overpopulated world the realities of scarcity mandates the enforcement of harsh measures. Hardin adopted a Hobbesian premise: left to their own devices self-interested human beings cannot be counted upon to act responsibly, especially in a society that promotes individual rights and freedoms. Only a Leviathan, or a managerial elite overseeing a tightly regulated market, could prevent the abuse of our environmental heritage. He states:

> If humanity were not infected with the poison of envy ...we could rely on individualistic voluntarism instead of laws. But we are descended from an unbroken line of envious ancestors, and it would be unwise to assume that we are different.... 'Nice guys finish last' (1977, 129).

What was needed, therefore, was not more but less freedom, for freedom must always be weighed carefully against the constraints of necessity.

Radical greens of the time were decidedly uncomfortable with such views. They felt caught in the middle of two extremes they equally refuted. On the one hand, there had developed an unpalatable authoritarian response to environmental degradation and, on the other, there was an equally unsatisfactory mainstream one. The time was right to propose a third. Despite 'the end of the world is nigh' gloom of the 1970s, many in the green movement instead took another detour. They began advancing value change as the key to arresting degradation. These radical greens commenced appraising in earnest the dominant values that underpinned environmental decay in an effort to challenge and substitute them. For some, anthropocentrism was identified as key, and became one of the main values to be censured. Others identified hierarchy and domination more broadly as key drivers of environmental ruin. All rejected both authoritarian and mainstream environmentalism because neither identified the paradigmatic values that drove degradation. In seeking to contain the negative impacts of industrial capitalism rather than to seriously transform it, these radicals believed mainstream reformists simply tinkered at the edges of any kind of meaningful change.

Mainstream environmentalism presented a reformist face. 'Mainstreamers' were essentially reformists who sought some lifestyle and behavioural change, in combination with the 'pressure group politics founded on the liberal assumption of a pluralist democracy' (Pepper 1993, 52). They did not reject capitalism, arguing instead for more appropriate small-scale versions of it. Attitudes towards government and the state varied, with some claiming a stronger interventionist and

regulatory role for the state, and others promoting more market autonomy. By and large, mainstream environmentalists embraced anthropo-
centrist, technocentrist and hierarchic values even as they injected
caution into how these values were interpreted and applied. O'Riordan's
original distinction between the radicals' ecocentrism and the mainstream's technocentrism remains apt, and captures well the different
positions promoted by each:

> [E]cocentrism....provides a natural morality – a set of rules ... based
> upon the limits and obligations imposed by natural ecosystems....
> Second, it talks of limits... and hence influence[s] the compass of
> 'progress'. Third, it talks in ecosystem metaphors of permanence
> and stability, diversity, creativity, homeostasis... Fourth, it raises
> questions about ends and means, particularly the nature of demo-
> cracy, participation... the distribution of political power and eco-
> nomic wealth, and the importance of personal responsibility....
> Finally, ecocentrism preaches the virtues of self-reliance and self-
> sufficiency (1976, 10–11).

By contrast:

> [Technocentrism] is identified by rationality, the 'objective' ap-
> praisal of means to achieve given goals, by managerial efficiency,
> the application of organisational and productive techniques that
> produce the most for the least effort, and by a sense of optimism
> and faith in the ability of man [sic] to understand and control phys-
> ical, biological, and social processes for the benefit of present and
> future generations (1976, 11)

Ecocentrism was very active during this phase of environmentalism.
In the process, it established environmental ethics as a respectable new
discipline in its own right. This was assisted by the national and
geographical focus of different environmentalists. In countries such as
North America and Australasia there developed a focus on wilderness
preservation that lent itself well to environmental ethics and the
extension of moral standing to entities other than humans. Also influential was the emerging animal rights discourse. In extending the
notion of sentience to animals (Singer 1975) and in inscribing them
with rights (Regan 1983) the matrix of moral considerability, even
within a liberal utilitarianism, was significantly widened. But it was
biocentrism that extended moral considerability furthest. In develop-

ing the notions of intrinsic value and biotic egalitarianism and in applying them to nature as a whole, deep ecologists lay the ground-work for the highly influential, if controversial, discourse of bio-centrism – later adapted to ecocentrism.

While anarchists were circumspect about ecocentrism, they took heart from the anarchical turn that radical ecology was now beginning to assume. In the early 1980s, adopting a mixture of ecocentric and anarchical sensibilities despite their party status, the German Greens pioneered a politics based on 'the four pillars' of environmentalism. These were ecology, social responsibility, grassroots democracy and non-violence; pillars underpinned by the principles of decentraliza-tion, post-patriarchalism and spirituality (Spretnak & Capra 1985, 3). Viewing the ecological crisis as one rooted in the 'spiritual impoverish-ment' of industrial society sat particularly well with an 'ecologising' anarchism that promoted awareness of 'our embeddedness in nature and the interconnected character of all phenomena' (Spretnak & Capra 1985, 50–1). At this point the fault lines between deep ecology and anarchism's social ecology had not yet ruptured.

While early ecologism may have claimed to have transcended the left-right spectrum by being 'neither left nor right but green', it became increasingly clear that it contained 'its own internal spectrum of debate' and its own competing 'political wings' (Eckersley 1992, 8). It had not 'transcended' the traditional spectrum altogether, even if it is difficult to position biocentrism on it. It has certainly introduced some new and distinctive considerations in the traditional matrix of 'who gets what, when, and how'. Many radical ecology schools still place traditional leftist concerns at their centre of their critiques. Questions of distribution, inclusion and justice dominate for most schools. But as Dryzek points out, at times 'the radically old and the radically new are combined in creative fashion' (2005, 183). We see this creative fusion in some of the radical ecology schools we now turn to.

Before we do, it is important to note that anarchist sensibilities have infused radical ecology both directly and indirectly. There are direct eco-anarchist schools – with Bookchin's social ecology prominent among them. Anarchist ideas have also infused other schools less directly – the theme that occupies us here. We highlight the penetra-tion of anarchist ideas in the discourses of deep ecology, eco-socialism/ eco-Marxism and eco-feminism, even when there is no 'formal' embrace of its ideology. Illustrating well how anarchical sensibilities have been incorporated in radical ecology more broadly, a noted eco-feminist comments:

Radical ecology emerges from a sense of crisis in the industrialised world. It acts on a new perception that the domination of nature entails the domination of human beings along lines of race, class, and gender. Radical ecology confronts the illusion that people are free to exploit nature and to move in society at the expense of others, with a new consciousness of our responsibilities to the rest of nature and to other humans. It seeks a new ethic of the nurture of nature and the nurture of people. It empowers people to make changes in the world consistent with a new social vision and a new ethic (Merchant 1992, 1).

This statement highlights some of radical ecology's key ingredients: the starting point of a world in moral and ecological crisis; an explicit challenge to domination and hierarchy in all its guises; the fashioning of a new environmental ethic; and the role of visionary politics in changing the world. Radical ecology's most distinctive feature is its 'social ecology'. The ecological relationship between nature and humanity that it uncovers and promotes distinguishes it from both mainstream environmentalism and traditional ideology as a whole (see Carter 1999).

Deep ecology

Deep ecology, or ecocentrism, wants to transform the relationship between humanity and nature, emphasizing interconnectedness rather than separation. It challenges an anthropocentrism that places human beings at the top of a hierarchy of value, with nature relegated to a freely available resource for human use. It suggests that the ecological crisis is rooted in a fundamental crisis of values. What is required is a new morality that takes into account ecological, as well as social and political, obligation. Instead of anthropocentrism, they propose a 'biotic egalitarianism' or 'biocentrism' that attributes equal value to all entities – human and non-human. For Deep Ecologists nature has intrinsic rather than simply instrumental value and its intrinsic value is sufficient to warrant its protection. The relegation to nature of the value it deserves, and the awareness of humanity's place in the natural order, requires a 'deep' consciousness change. Only through a radical moral extensionism that renders the non-human world valuable in and of itself, can ecological and moral crisis be averted. Deep ecology's main challenge is to anthropocentrism – a radically extended notion of hierarchy.

The principle of biotic egalitarianism, or biocentrism, which it opposes to anthropocentrism underpins deep ecology. As early as 1967 a critic of anthropocentrism suggested that 'what people do about their ecology depends on what they think about themselves in relation to things around them' (White Jr 1973, 24). Furthermore, this 'logic of domination' is traced to a Christian, scientific/technocratic rationality where 'despite Darwin, we are not, in our hearts, part of the natural process; rather, we are 'superior to nature, contemptuous of it, willing to use it for our slightest whim' (1973, 28). If human beings continue to view nature instrumentally, the logical conclusion becomes one where 'we will continue to have a worsening ecologic crisis until we reject the Christian axiom that nature has no reason for existence save to serve man' (1973, 29). Moderated since its early formulation – to account for some of the misanthropic and political quandaries it unwittingly unleashed – deep ecology still demands significant consciousness change and a new way of 'being' in the world. Its 'organismic democracy' emphasizes the interrelatedness of all living phenomena, attributing 'equal importance to every component of the interlinked web of nature' (Merchant 1992, 86). The operation of this 'organismic democracy' demands a new psychology and spirituality of self so that there is 'a total intermingling of person with planet' (Merchant 1992, 86). The self-interested and atomistic individualism promoted by both classic liberalism and modernism is rejected for more relational conceptions of self. Deep ecology thus addresses the spiritual impoverishment of industrial society:

> [D]eep ecology is spiritual in its very essence. It is a world view that is supported by modern science but is rooted in a perception of reality that goes beyond the scientific framework to a subtle awareness of the oneness of all life, the interdependence of its multiple manifestations... When the concept of the human spirit is understood in this sense, as the mode of consciousness in which the individual feels connected to the cosmos as a whole, the full meaning of deep ecology is indeed spiritual (Spretnak & Capra 1985, 50).

Deep ecology draws from a range of philosophical traditions including Spinoza, Leopold and, more recently, Arne Naess. Naess is considered the 'father' of deep ecology, and one of the first philosophers to defend a 'deep' ecological position against a 'shallow' one. 'Shallow ecology', like mainstream environmentalism, simply seeks to reform industrialism to take account of environmental degradation's impact on people.

Instead, deep ecology uncovers 'deeper concerns which touch upon the principles of diversity, complexity, autonomy, decentralization, symbiosis, egalitarianism, and classlessness' (Naess 1973, 95). The capacity to ask deeper questions rests on the shift from a dominant rationalist paradigm to one based on a spiritual and intuitive 'way of knowing'. For Naess, 'the norms and tendencies of the Deep Ecology movement are not derived from ecology by logic or induction' but through an intuitive process that allows for 'Self-realization' (1973, 98). A Self-realized human being experiences, both emotionally and intuitively, a sense of their connection to 'something greater than their ego... greater than their attributes as an individual' (Devall & Sessions 1985, 76). This relational intuition enables the lower case self to transform to a larger Self intimately interconnected with nature: a self-in-Self or the individual as a symbiotic part of nature. This extended identification with nature means that to exploit or destroy it is to exploit and destroy a part of oneself. When Deep ecologists claim that 'no-one is saved until we are all saved', they include 'all humans, whales, grizzly bears, whole rainforest ecosystems, mountains [and] rivers' (Devall & Sessions 1985, 67).

Another way of theorizing this deep ecology position is to claim, as Aldo Leopold (1968) does, that human beings are 'plain members of the biotic community, not lord and master over all other species'. To seek reform of a system whose very premise is flawed, as mainstream environmentalists do, is simply to invite failure. Short of challenging the anthropocentric roots of environmental ruin, and advancing the spiritual and behavioural changes required to address it, degradation of ecological systems will continue unimpeded. But while they chastise mainstream environmentalists, deep ecologists do not necessarily speak for all radical ecologists. Deep ecology's radical roots are more philosophical than political, although as interpreted by Earth First! they take on a decidedly stronger – and anarchical – politics.

Importantly, not all deep ecologists are automatically anti-capitalist, even if ecocentrists like Earth First! are hostile to it. It is often for their refusal to directly condemn capitalism that other radicals condemn deep ecologists. The ecocentric objective is not the overthrow of capitalist *per se*. Their objection is more to industrialism than capitalism, even if capitalism has finessed industrialism considerably. Many seek to 'ecologise' capitalism by instilling 'some order into the present conceptual chaos' and 'allowing economists to put their models into an ecological context' (Spretnak & Capra 1985, 79). A key criticism of economic growth lies in its marginalization of ecological criteria: 'We don't say every form of growth has to be rejected. We reject it only as long as this functional connection [between growth and ecological destruction] continues to exist' (Muller in Spretnak & Capra 1985, 85). Deep ecologists generally support the use of 'soft' and 'appropriate' technologies and make ecological sustainability the bottom line. This is often measured in terms of maintaining the integrity of the natural environment by not putting negative pressure on its natural carrying capacity.

Overall, however, deep ecocentrism defends its position through a new environmental ethic that it opposes to the dominant, human-centred ethic of modernism. Concepts such as interdependence, complementarity, holism and interrelatedness are now central to the articulation of an ecocentric position. Capra's (1983) contention that the spiritual and physical health of human beings is dependent on an integrated ecosystem, and that human beings are just one, albeit important, link in this ecosystemic chain, is a view that is still generally accepted among deep ecologists. Most deep ecologists are therefore concerned with articulating a theory of value that incorporates the intrinsic worth of both human and non-human entities. In short, they wish to transform the nature/culture relationship on a number of levels, including the economy.

Deep ecology, and translations such as 'transpersonal ecology' requires, then, a new psychological and spiritual approach that demands human beings know who they are before they can change what they do (Fox 1990). An alienation imposed by hierarchical and anthropocentric thinking blinds humanity to the fact that they are a part of nature and that only by protecting nature can they protect themselves. A transpersonal ecology changes this by reactivating humanity's personal, ontological and cosmological identification with nature (Fox 1990, 249–58). A transpersonal ethic based on extended identification with the nature shuns the instrumentalist and human-centred sources of ecological havoc. It shares much with Mathews' (1991; 1995) conception of the ecological self – a relational self she opposes to the damaging 'separate self' of liberalism. Ecological restoration thus requires a new relational way of 'being' in the world, a being that rids itself of both physical and 'mind' pollution.

Deep ecology has introduced a novel and influential ethic into the discourse of environmentalism. While this ethic has been vigorously censured, it continues to define important components of environmental politics, as we observe in our discussion of Earth First! in Chapter 8. Most other radical ecology schools have articulated their theories in at least some relation to it. Many reject a biocentrism that they claim made little sense ecologically or politically, while others incorporate a tempered ecocentrism. Some articulate their objections shrilly, with the vitriolic debates between the deep and social ecologists in the 1980s testimony to this. It is deep ecology's biocentrism that irks most. The targeting of 'humanity' as ecological culprit is considered at best politically shortsighted, and at worst misanthropic and ethnocentric. Laying the blame on an amorphous 'people' ignores the different class and race experience of environmental degradation, as well as the structural underpinnings of inequality. These critics charge that consciousness change does not take place in a vacuum. Eco-socialists argue that economic and political conditions shape environmental attitudes far more than psychology or spirituality – which are in any case post-material values pertinent only to the well resourced middle classes. Eco-feminists points out that deep ecology is 'both individualistic (failing to provide a framework for change which can look beyond the individual) and psychologistic (neglecting factors beyond psychology)' (Plumwood 1993, 17). In short, while deep ecology may rail against the operation of hierarchy against nature, it can overlook its operation within humanity.

While deep ecology does not claim any direct anarchist roots, it identifies hierarchy as culprit in environmental ruin. Its conception of hierarchy may be novel and applied relatively exclusively, but it highlights a hierarchy of value wielded by the powerful over the powerless – with inert nature as the most subjugated. Some deep ecologists continue to champion personal change as the way forward. Others such as Earth First! unashamedly trumpets its direct and combative defence of the 'powerless' earth. Someone, they claim, has to defend an inert nature against the hierarchs who would devastate it. Most radical greens, including anarchists, have listened attentively if cautiously to the claims of deep ecology. Earth First! goes considerably further, marrying its deep ecology with an anarchist politics. But most radical ecologists have embraced at least some of its ideas, particularly of an ecocentrism that extends the notion of community to also encompass nature. Since nature cannot 'speak', they see no reason why human advocates should not 'speak' on its behalf (see Eckersley 1999). But in seeking the more active participation from all members of its extended communities, ecocentrism promotes a 'deeper' democracy that has widespread appeal.

Eco-feminism

Eco-feminism stresses a connection between environmental ruin and patriarchy, arguing that there are direct links between the domination of women and the domination of nature. A 'logic of domination' renders both women and nature 'other' and hence inferior. With its roots in patriarchy, this 'logic of domination' then acts as the template for all other forms of domination, including the domination of nature. For many eco-feminists, environmentalism is essentially a feminist issue, for by challenging the patriarchal logic of domination not only women, but nature, are liberated. In The Death of Nature (1980) Carolyn Merchant argues that the seeds of a modernist patriarchy lay in the scientific revolution of the 17th and 18th centuries. This revolution transformed both nature and women into resources under the control of men. In transforming nature 'into a machine to be controlled and repaired by men', both women and nature 'were subordinated to male-defined purposes' (Merchant 1987, 18). An emerging capitalist industrialism then finessed the devaluation of women and nature. Neither women nor nature can be freed without confronting the underpinning 'logic' of domination:

Since the exploitation of nature is bound to social processes that oppress people, and since the logic of these systems of domination is modelled on the logic of male domination, neither nature nor women will be liberated without an explicit confrontation with these structures of male domination (Young 1983, 175).

Eco-feminists thus identify hierarchy in the form of a patriarchal logic of domination as culprit in environmental ruin. This logic of domination is buttressed by two key values. The first is an anthropocentric one that privileges human beings over nature. The second is an andocentric one that privileges men over women. Eco-feminists thus go beyond a broadly conceived anthropocentrism since it can mask a dominant androcentric practice. Since 'women are identified with nature and the realm of the physical' and men are identified with the realm of the 'human' and the 'mental', and since 'whatever is identified with nature and the realm of the physical is inferior to whatever is identified with the human and the realm of the mental' men are hence justified in subordinating both women and nature (Warren 1990, 130).

There is nonetheless much internal disagreement among eco-feminists over how best to conceptualize the patterns of domination that link women and nature. All eco-feminists reject a status quo liberal feminism, however. They allege that liberal feminism overlooks the broader reaches of domination, seeking instead to win women equal access to the fruits of hierarchy and privilege that men enjoy.

Liberal feminism's conception of 'liberation' and 'empowerment' is defined in masculinist terms, rendering it toothless in the struggle for broader egalitarian and ecological goals. Social eco-feminists simply ask: 'What is the point of partaking equally in a system that is killing us all' (King 1990, 106). Liberal feminism is considered insufficiently analytical and insufficiently radical, ignoring the structural causes of exploitation and domination. It seeks to 'fit women somewhat uncritically into a masculine model of humanity and culture'; an 'uncritical equality' that simply demands 'equal admittance for women to a masculine-defined sphere and masculine institutions which are criticized only to the extent that they exclude women' (Plumwood 1992b, 11). If at all green, liberal feminists are shallow environmentalists, seeking limited environmental and gender reforms. In short, there is nothing ecological or radical about liberal feminism.

Cultural and social eco-feminism

Eco-feminism is often categorized into cultural eco-feminism and social eco-feminism. Cultural eco-feminism celebrates the special relationship that women are deemed to have with nature. This relationship has both a biological and spiritual component. Women's intimate link to nature stem from shared biological and reproductive capacities, an affinity necessarily denied men. Women's reproductive functions equip them with the qualities of care, nurture and emotion. These qualities enable women a greater identification with nature and hence a stronger empathy with nature's violation. Rather than disentangling the nature/women connection, cultural eco-feminists celebrate it. Many cultural eco-feminists pin their hopes for ecological/cultural renewal on the very qualities that women share with nature. Masculine values emphasize separation and domination (of both women and nature) instead of connectedness and nurturance – the very qualities required to rescue the natural world from destruction. Cultural eco-feminists thus propose an alternative 'women's culture' or 'authentic female mind' which is 'based on revaluing, celebrating and defending what patriarchy has devalued, including the feminine, non-human nature, the body and the emotions' (Plumwood 1992a, 10). These qualities thus need to be rescued from the scrapheap that a patriarchal society has relegated them.

For many cultural eco-feminists, release from a dominant patriarchy comes through feminist spirituality and feminist rituals. Goddess symbology and pagan ritualism helps celebrate the close links between women and nature. 'Goddess worship' celebrated life affirming procreation and fertility rites but the emergent patriarchal culture debased such qualities. As one Goddess historian argues, 'for millenia... prehistoric societies worshipped the Goddess of nature and spirituality, our great Mother, the giver of life and creator of all', and 'even more fascinating' these societies were renowned for their gender and ecological harmony, qualities contemporary humanity is now seeking (Eisler 1990, 23). Now, women are associated with nature, and men with culture, where nature equals 'lesser' and is the domain of women and the natural world, and culture equals 'better' and is the domain of men (Merchant 1992, 192). For cultural eco-feminists, biology helps define human nature and reproductive biology helps characterize female nature, as well as nature itself. A masculinist culture rests on a scientistic, technocratic and rationalist hierarchy of value. By implication this devalues both women and nature. The revaluation of nature requires a social upheaval that celebrates 'women's biology and Nature... as sources of female power' (Merchant 1992, 192).

Social eco-feminists reject such theorization as biological determinism. They charge that a cultural eco-feminism so conceived creates a universalist and essentialist conception of female nature. Its apolitical temperament overlooks the cultural and socio-economic differences between women – criticism they have also levelled at deep ecology, with which cultural eco-feminism is often associated. Social eco-feminists also highlight the link between the domination of women and of nature but incorporate social, political and economic factors more directly in the matrix of domination. They avoid the biological and cultural reductionism of cultural eco-feminism, and do not view patriarchy as the template of domination. They identify patriarchy as formidable, but avoid ascribing it the status of 'template of oppression'. Patriarchy is an important component of domination, but domination is also far more complex than this. While acknowledging the link between the domination of women and the domination of nature, social eco-feminists are careful to avoid the biological and cultural essentialism of their eco-feminist counterparts. Nor do they subscribe to patriarchal reductionism. Rather they view 'the threads of gender as interwoven with those of class, race and species' (Plumwood 1992a, 10).

Not unexpectedly, socialism and anarchism inform social eco-feminism. Socialist eco-feminists compensate for Marx's omission of gender and nature in the matrix of domination, however. They adapt production of labour to more appropriately include its reproductive elements. They find in Engels a more sympathetic treatment of the gendered relations of production. They argue that just as man's productive labour forms the material basis of life and is appropriated by the ruling class, so too is women's reproductive labour. Women are thus condemned to a two-pronged oppression. They suffer the general subjugation of the working class of which they are a part, and to the additional consigning of their reproductive capacities to producing capitalism's future labour force. By 'elaborating on Engels' fundamental insights women's roles in production, reproduction, and ecology can become the starting point for a socialist ecofeminist analysis' (Merchant 1992, 196–7), which it did for many socialist eco-feminists.

Social eco-feminism includes many socialist eco-feminists who continue to identify class as the major source of inequality and the pivot on which oppression revolves. They highlight a strong class component to environmental degradation, arguing that the poor, which includes many women, are more likely to suffer the consequences of environmental problems than the better resourced. They are thus highly critical of a deep ecology that seeks consciousness change in a seeming political vacuum and of cultural eco-feminist positions that ignore the structures of inequality. For socialist eco-feminists meaningful social change demands the transformation of capitalist structures of domination. Only then can the previously oppressed, including women and nature, prosper. The prominent Indian eco-feminist Vandana Shiva (1988) observes that the link between destruction of nature and the destruction of women's livelihoods manifests both materially and ideologically. Attributed limited cultural and gender respect, and often responsible for the material livelihood of their families, when nature is destroyed so too are the means for these women's survival. At stake, especially in the vulnerable south, are women's capacity for staying alive. Shiva has described globalization as a form of 'environmental apartheid': it restructures 'control over resources in such a way that the natural resources of the poor are systematically taken over by the rich, and the pollution of the rich is systematically dumped on the poor' (1999, 53). Under these conditions it is primarily women, and their families, who suffer.

Anarchist social eco-feminists agree with their socialist colleagues, but stop short of assigning the state any redemptive role. Prominent anarcho-feminist and social ecologist Janet Biehl contends that social eco-feminism 'accepts the basic tenet of social ecology, that the idea of dominating nature stems from the domination of human by human' (Biehl in Merchant 1992, 194). Yet anarcho-feminists demur from too strong an attachment to a class analysis, preferring hierarchy as main culprit, even if class is a highly charged form of hierarchy. But anarcho-feminists reserve most of their criticisms for cultural eco-feminism. They see 'goddess worship' as at best another 'trendy' religion for the bored affluent. At worst, like most religions, it assembles a new form of authoritarianism and hierarchy. But spiritual eco-feminists do not let such criticisms go undefended. They charge that while their socialist counterparts 'have addressed one of the three forms of domination of nature – domination between persons – they have not seriously attended to the domination of non-human nature, nor to the domination of inner nature' (King 1990, 114).

Social eco-feminists view gender as a social construction rather than a biological determinant. They are more comfortable with a continuum rather than polarization of masculine and feminine qualities. They seek to server rather than celebrate the nature/culture link. They view the use of women's alleged closeness to nature to revalue female culture as both reductionist and essentialist, ignoring the other factors that shape women's identity. The reductionist casting of women as 'good' consequently casts all men as 'evil', overlooking the fact that 'history, power, women, and nature are all a lot more complicated than that' (King 1990, 111). A dominant value hierarchy positions human beings at the top, and has historically divided this echelon between men (higher) and women (lower). But this does not preclude a hierarchy between women, nor a female one towards nature. A broad anthropocentrism might privilege androcentrism over 'femocentrism' but anthropocentrism includes women, and also privileges their instrumental rights over nature. As human beings they are still 'above', even if one hierarchical step removed. Even if there is a female 'association with nature' this does not automatically translate to an 'identification with nature': many women 'have been conceived, and conceive of themselves as dominators within the logic of domination – as above nature, and/or as above other members of the human species' (Cuomo 1992, 356). Critics thus find in cultural eco-feminism the same 'soft' anthropocentrism of deep ecology (Luke 1988). Others interpret its reference to nature as 'Mother Nature', 'Earth Mother' or 'Earth Goddess' as anthropomorphism at its worst; 'it is the old game of projection upon nature from the human for human needs' (Dodson Gray 1989, 275). Still others defend the use of certain gendered terms as political strategy. Commenting on her involvement in the anti-globalization movement as a 'witch, pagan, feminist and anarchist' Starhawk highlights the power of discourse in transforming the world:

To move beyond our usual categories of thought, we need words that shock and confuse and shake up our usual thinking ... The use of the term 'Witch' to describe myself was a political and spiritual choice I made long ago. Political because I felt that to challenge the deep misogyny in our culture, the ingrained fear of women's power ... we had to make visible those underlying thought structures and challenge them, and the word 'Witch' does that (2003, 264–5).

Overall, eco-feminism constructs compelling links between the domination of women and the domination of nature. Their central insight that the domination of nature is enacted through the same logic that controls an array of other social arrangements, including patriarchy, capitalism, racism and instrumentalism, remains an important one. This insight helps mark out eco-feminism's anarchical temperament. The finer detail may lie in the discourses of androcentrism that different eco-feminists produce but, like deep ecology, eco-feminism too identifies hierarchy, broadly defined, as central to the practice of domination.

Eco-socialism

Eco-socialism rejects deep ecology's biocentric ethic and defends a humanist politics. Its main objection is to a capitalist industrialism rather than to industrialism *per se*. To the litany of abuses capitalism perpetrates, they now add the degradation of nature and the inequitable distribution of this degradation to the most vulnerable. Their main concern is thus the equitable distribution of resources which now includes equitable distribution of environmental quality. Their focus is on the strong link between environmental degradation and the class experiences of it, with the poor, people of colour and women more likely to suffer the consequences of degradation. While both seek to conserve nature, debates between eco-socialists and eco-anarchists reflect long standing historical conflicts. Eco-socialism continues to look to an 'enabling' state as a partner in social and environmental change. But unlike deep ecologists, most eco-socialists are proud humanists who do not necessarily resile from anthropocentrism. Eco-socialism thus opposes itself to both eco-anarchism and deep ecology. The 'red-green' debate encapsulates some of these quarrels.

Like eco-feminism, eco-socialism incorporates a spectrum of views and positions. We can group them into two broad categories. The first is eco-Marxism which largely upholds a traditional Marxist analysis in its accommodation with environmentalism. The second is a more tempered eco-socialism that goes considerably further in its efforts to green Marxism. This broad division is nonetheless not so straightforward and encapsulates a diverse range of positions, including orthodox eco-Marxists, humanist eco-Marxists, critical theorists, post-Marxists and welfare ecologists (Eckersley 1992). Orthodox eco-Marxists remain relatively 'unreconstructed' Marxists who continue to utilize, defend and expand on its basic tenets, especially of the 'mature' Marx. They generally attribute environmental problems to the capitalism's relations of production, even as they expand Marxist analytical categories considerably.

A prominent orthodox eco-Marxist, James O'Connor, has linked environmental degradation to the logic of capitalist economy in an influential and innovative way. O'Connor builds on Marx's theory of economic crisis – that is, that the logic of capitalism leads it to generate its own contradictions, or barriers, to economic expansion. These contradictions manifest between capitalist productive forces and production relations, or 'between the production and realization of value and surplus value, or between the production and circulation of capital' (O'Connor 1988, 15–16). In this way, the 'limit to capital is capital itself' (O'Connor 1987, 50). For O'Connor, such conditions today include the 'external physical conditions' of environmental pressures; and its contradiction becomes the 'agency of social transformation', or new social movements such as environmentalism (1988, 17). These crisis tendencies are exacerbated by the globalization of 20th century capitalism, especially 'the growth of finance capital and monopoly capital and new roles for the political system and state bureaucracy in the process of capital accumulation especially with regard to capital restructuring' (1987, 51–2).

For O'Connor, contemporary capitalism has thus expanded its tendency to crisis through a second contradiction: environmental degradation. He contends that while Marxism's conventional focus on the exploitation of labour was critical, he overlooked the equally important exploitation of nature – or exploitation of the resources on which capitalist development depended. Hence, 'just as the labor movement forced capital to move from a mode of absolute surplus value to one of relative surplus ... so the green movement may be forcing capital to end its primitive exploitation' of nature (1994, 158). This means that there may be more than one 'path to socialism', the first through the traditional labour movement, the second *via* the new social movements. Appealing to green (and feminist) socialists alike, O'Connor contends that environmentalism may represent a 'parallel path to socialism', making Marx 'not so much wrong as he was half-right' (1988, 27).

Humanist eco-Marxists interpret Marxism in an 'ecologically-friendly' light by building on its implicit, if limited, ecological dimension. They challenge Marx's productivist optimism, arguing that in relying on the exploitation of natural resources, productivism can be inherently dangerous. Deriving more of their Marxism from the alienation discourses of the 'young' Marx, they seek to transcend the alienated relationship between humanity and an external nature. They revise what is generally perceived as a 'mastery of nature' discourse in Marxism, even as they hold on to a tempered humanism. They acknowledge that Marx's 'nature philosophy' is difficult to discern, not least because of the seemingly contradictory views of the 'young' Marx as opposed to the 'mature' one. Few defenders of Marx's nature philosophy would attempt to locate a notion of intrinsic value in his works, yet would defend him from a charge of unadulterated human chauvinism. They take heart from Marx's own observations: 'Estranged labour turns... Man's species-being, both nature and his spiritual species property, into a being alien to him... It estranges man's own body from him, as it does external nature and his spiritual essence, his human being' (Marx in Tucker 1978, 77). Defenders of Marx's ecology thus charge that it is the process of alienation itself that turns humanity against nature, and that reinforces the nature/culture divide. In emphasizing an interdependent rather than dominatory relationship between humanity and nature, a defender of the humanist position argues that Marx articulated a 'dialectical relationship' between humanity and nature, one where 'man (sic) transforms it and is therefore transformed' (Parsons 1977, xi).

Critical theorists go further, challenging key Marxist tenets such as its instrumentalism, scientism and historical materialism. While they did not specifically develop their own green theory they lay the groundwork, as we saw, for situating the practice of domination well beyond class. In the process, they helped highlight the link between the domination of nature and the domination of human by human, a link that would greatly influence Bookchin in his formulation of social ecology. Post-Marxist eco-socialists are considered the 'greenest' both in their acknowledgment, if not blanket endorsement, of the intrinsic rights discourse and in their willingness to temper the Marxist productivist ethos with the realities of ecological limits to growth.

Eco-socialist principles

Eco-socialists of all kinds acknowledge, in various degrees, the challenge that environmental degradation has imposed on the traditional Marxist discourse. While some may have been originally antagonistic, especially those in the 'old' labour movement, they are now more prepared to work with the new social movements, even if this accommodation can remain fragile. In the early 21st century, however, and especially through the anti-globalization movement, there is a much stronger accommodation. As we saw in Chapter 2, the Seattle protest in 1999 was particularly notable for the oppositional unity between labour, unions and greens, as well as a plethora of others. But despite different emphases and interpretations, most eco-socialists subscribe to a number of key principles.

First, there is general recognition that environmental risk also lies in the process of industrial development, whether communist or capitalist, and that mitigation of such risk requires direct and precautionary responses. Socialism or communism, in themselves, are no automatic guarantee of environmental harmony. Soviet authoritarianism was no less repressive than the capitalist domination it sought to replace, including its repression of nature. There is now general cognisance that Soviet communism's environmental record was very bleak. Eco-socialism was thus forced to revisit the sensitive issue of development and economic growth, one that remains contentious today. While eco-socialism no longer supports unmitigated growth, it does not demur from its second key principle: a strong distributional focus. Of more concern than how wealth is created is its equitable distribution. This now includes ensuring that there is also equitable distribution of environmental quality and that environmental risks do not fall disproportionately on the shoulders of the poor and powerless. The eco-socialist critique of capitalism is nonetheless trained more towards capitalism's exploitative thrust – the relations of production – than towards its productivist thrust – the forces of production. An underpinning productivism continues to interpret what nature provides as resources to be utilized for growth, even if many eco-socialists champion sustainable or ecologically benign growth.

Third, in some form or other, class remains central in eco-socialist discourses. The eco-socialist focus on distributive justice impels them to continue linking environmental risks with the class experience of them. But most eco-socialists have expanded the contours of domination beyond class to also include the important categories of race and gender. The related environmental justice movement, for example, highlights the stark link between poverty and environmental degradation (Commission for Racial Justice 1987; Bullard 1990; Bryant & Mohai 1992; Goldman 1992; Westra & Wenz 1995). It notes that many 'economically impoverished communities and their inhabitants are exposed to greater health hazards in their homes, on the jobs, and in their neighbourhoods when compared to their more affluent counterparts' (Bullard 1997). Environmental concerns are thus social concerns and remain issues of maldistribution and inequitable power relations. What eco-socialists object to is not so much the damage perpetrated on nature, although they are deeply concerned by this, but the refusal by many greens, especially deep ecologists, to acknowledge that:

> ... it is accumulation of wealth and its concentration in fewer and fewer hands which creates the levels of poverty that shapes the lives of so many people on our planet... [and that]... it is poverty which forces people to place their own short-term interests above the long term interests of the Earth's ecology (Weston 1986, 5).

In short, eco-socialists decry the lack of a rigorous political analysis in, at least some forms of, ecologism. Class goes to the heart of this rigour and much eco-socialist criticism of other greens has centred around the issue of class. They charge deep ecologists in particular with championing a middle-class elitism that ignores the link between class and environmental risk. They claim that 'green-greens' (deep ecologists) 'need to move beyond their ecological determinism'; and green-greens in turn charge that the 'red-greens' (eco-socialists) 'need to move beyond the traditional parochialism of trade unionism and away from the old-fashioned ideas of centrist political control and perpetual industrial expansion' (Weston 1986, 5). Nonetheless, for eco-socialists it is only through a creative synthesis with the experiences of working-class peoples – both nationally and internationally – that the environmental movement can shake off its reputation as a limited middle-class phenomenon. However there is not as yet an eco-socialist consensus on the best strategic way forward for green politics. Some promote the institutional route of greened social democratic parties or dedicated green parties, others emphasize the role of civil society and the grassroots.

Finally, eco-socialists hang on to an unashamed, albeit tempered, anthropocentrism. Many promote a human welfare approach that they oppose to the idea of intrinsic value and the rights of nature. They believe that a strong environmentalism that is respectful of nature can flow from humanity's realization that it is in their interests to protect it. They assert that a human interest in protecting nature is not the same as overlooking its intrinsic grandeur. In short, in 'just the same way [that] it is legitimate for elephants to prefer what is in the interests of elephants' it is legitimate for humanity to prefer what is in its own interests – a balanced and unspoiled natural world (Pepper 1993, 222). Such an outcome benefits both nature and people, while also providing for 'human spiritual welfare' (1993, 222). The primary goals of ecological restitution remain human-centred ones – that is, the extension of justice measures to all members of the human community. But this privileging of human concerns does not automatically preclude respect for the integrity of non-human nature.

Overall, eco-socialism continues to emphasize 'people's collective power as producers' and hence situates both labour and the environmental crisis at the centre of the capital-labour relationship (Pepper 1993, 233–4). While eco-anarchists too find much to condemn in the capital-labour relationship, it represents for them only a part of the environmentalist story, with hierarchy the main analytical thread. Institutional and statist solutions – whether green parties, sustainable development or stronger social democracy – are considered equally wrong-headed. A focus on such strategies put an even greater distance between eco-socialists and eco-anarchists. Already circumspect about eco-socialism's class focus and its productivist ethic, eco-socialism's continued championing of the institutional route represents, as always, the biggest problem for eco-anarchists. But eco-socialism has moved a considerable way towards acknowledging the ecological limitations of productivism and industrialism, the political limitations of vanguards and the operation of domination beyond class. Anarchists lay claim to having helped foster these insights.

Conclusion

This chapter has explored a number of radical ecology schools and identified their anarchical influences. Like the other schools, eco-anarchism is also a broad position that incorporates a number of different 'sub-schools', the main ones being social ecology, bio-regionalism and primitivism. But, highlighting the difficulties of classification, primitivism is often included under the umbrella of social ecology, albeit a very different one to Bookchin's social ecology; and bio-regionalism has strong deep ecology roots, as does primitivism. Bookchin's social ecology remains one of the main eco-anarchist schools today, and is the subject of Chapter 6. Overall, however, anarchism has exerted considerable influence in shaping many facets of radical ecology, with most schools identifying hierarchy in some form or other as central to the operation of both social and ecological domination. And radical ecologists also embrace considerable aspects of an anarchist politics.

Radical ecology's 'prime' may have been two or three decades ago when a flurry of theoretical activity was devoted to the development of novel eco-political positions. But as an important part of the highly successful green movement, it helped shape the green values that have clearly penetrated contemporary oppositional politics. Most radical groups today incorporate at least some aspects of its discourses. There are also numerous eco-centrists, eco-socialists and eco-feminists in the anti-globalization movement, alongside eco- and post-ideological anarchists. As the corporate 'assault' on the environment continues apace, radical groups such as Earth First! and Reclaim the Streets are determined to resist it. They are equally determined to keep alive ecologism's radical heart.

This chapter completes the book's first part. This part has detailed the main theoretical and political developments that have helped shape anarchism's contemporary influence. Drawing on these influences and these themes, the next section explores the practice of post-ideological anarchism through a number of illustrative case studies.

Part II

Practising Contemporary Anarchism

5
The Politics of Zapatismo

Enough is enough! So declared the Zapatista National Liberation Army (EZLN) that emerged from the jungles of southern Mexico to occupy the town of San Cristobal de las Casas on New Year's Day 1994. On the same day that the North American Free Trade Agreement (NAFTA) was proclaimed, the Zapatistas declared NAFTA a 'death sentence for Indigenous people'. They decreed *ya basta* (enough!) to the relentless neo-liberal policies that further devastated the already meagre livelihoods of the Indigenous populations in Chiapas – one of Mexico's poorest states. Dressed in black ski masks, carrying an assortment of real and fake guns, and led by a female soldier, these masked Indigenous guerrillas occupied the Municipal Palace and erected their black EZLN flag there. A solitary masked figure with bandoliers across his chest then mounted the Palace balcony to declare war against the Mexican state:

> We are a product of five hundred years of struggle ... we have nothing, absolutely nothing, not even a roof over our heads, no land, no work, no health care, no food or education ... But today we say: ENOUGH IS ENOUGH! We are the inheritors of the true builders of our nation. We are millions, the dispossessed who call upon our brothers and sisters to join this struggle (Marcos 1994b, 13).

This *ladino*, or Spanish Mexican, who became the Zapatistas' spokesman, was later identified as Subcommandante Insurgente Marcos – the articulate, passionate and politically savvy 'postmodern' rebel who captured not only Mexico's but the world's heroic imagination (see Esteva & Suri Prakash 1998).

The Zapatistas assumed temporary control of several towns before being dispersed by the state's security forces in skirmishes that cost many guerrillas their lives. The EZLN's fire power was obviously no match for the state's military and police, and their insurrection was quickly contained. Defeated in a battle they knew they had little hope of winning, the EZLN nonetheless counted some significant wins. They not only trained Mexican attention on the sorry plight of its Indigenous, but helped consolidate global resistance to neo-liberal globalization. As the Indigenous poor of Chiapas, the Zapatistas functioned simultaneously on three levels: the local, national and global. Since the conditions of the Chiapas region was a product of its national politics, and since national responses were in turn conditioned by global politics, the Zapatistas recognized that only by a simultaneous three pronged challenge could they sustain their resistance. On a strategic level, national and global visibility helped protect their communities from state reprisals. Their objective was not in any case the capture of state power or the instigation of an on-going conflict with the Mexican state. Rather, 'the task of an armed movement should be to present the problem, and then step aside' (Marcos in Weis 2003, 14).

In a little over a decade, the Zapatistas have helped define the character of 21st century dissent. Their 1994 uprising seized global attention and stirred the impetus for radical change. It was an important turning point for oppositional politics worldwide. In the light of globalization's impact on Mexico's Indigenous poor, the Zapatistas made it plain that they had reached the end of their political tether. Their direct and dramatic rebellion against globalization's ills helped drive the hostility to neo-liberal globalization that would crystallize in the anti-globalization movement (AGM). In doing so, they launched an ideological and 'tactical template' that would be embraced by this movement. *Ya basta* soon symbolized both the temperament and vigour of global dissent and was an increasingly visible banner in anti-globalization actions throughout the world. Two years later the Zapatistas' Subcommandante Marcos issued an invitation to activist groups and social movements across the globe to participate in the First Intercontinental Meeting for Humanity and Against Neo-liberalism (the *Encuentro*). The planned location of the arduous Mexican jungle was expected to dampen participation considerably. Instead, on 27 July 1996, 3000 activists from forty countries in five continents settled into the hand-crafted conference centres to begin a global counter-insurgency against a common adversary: neo-liberal globalization (Notes from Nowhere 2003, 34). This global coordination of dissent helped launch Zapatismo as an inspired political philosophy for the 21st century.

But the question remains: of the many Indigenous resistance movements across the globe, why did the Zaptatistas strike such a global
chord? An important part of the answer lies in Zapatismo's oppositional
character and its distinctive political philosophy. Another part lies in the
sophisticated delivery of its message. Through his compelling articulation of Zapatismo, Marcos turned dialogue into a powerful political tool
that transformed 'words' into 'weapons'. The Zapatistas are an 'army'
that dialogues rather than fights and an army that fires words of rebuke
against soldiers in Mexican military bases instead of bullets. As a resistance movement they are equally comfortable with poetics as with
protest – indeed with poetics *as* protest. Their inspirational words were
then launched as communiqués through the internet to the global
community. As a 'postmodern' rebel, Marcos was not averse to utilizing
modern technological tools to disseminate the ideas of Zapatismo.
There are now over 45,000 Zapatista-related websites globally and
Zapatista communiqués have been translated into 14 languages
(Burgmann 2003, 293). The communication technologies employed to
cyber-launch their political dialogue has ensured them global visibility.
Their consequent global 'celebrity' not only strengthens their political
message, but also protects them from political retribution within their
own borders. As a 'weapon', their words serve multiple functions.

The Zapatistas occupy a central presence in contemporary oppositional politics. Many AGM actors embrace them as inspirational
beacons and ideological allies. Marxists, socialists and anarchists all
seem to claim them as their own. The Zapatistas themselves refuse the
restrictions of ideology and decline to call themselves anything. This is
despite their indebtedness to Marxism, socialism, anarchism and liberation theology. While the Zapatistas refuse to name an ideological
identity, other political actors are not so reluctant. This includes anarchists who are heartened by what they identify as strong anarchical
impulses in Zapatismo. In exploring the character of Zapatismo, this
chapter stakes a large claim: that without necessarily being anarchists,
the Zapatistas represent the post-ideological anarchist position very
well. They draw flexibly and openly from anarchism, and a range of
other traditions, to fashion their own non-ideological 'brand' of political philosophy. The chapter first explores the rise of the Zapatistas,
situating them in the fraught Indian history from which they arose
and that helped 'grow' them. The main features of Zapatismo, particularly as articulated by Marcos, are then discussed. We finish by considering the links between Zapatismo and post-ideological anarchism.

Land struggles and the rise of the Zapatistas

The rise of the Zapatistas must be framed in the history of the Indigenous experience in Mexico. Two factors stand out here: first, the politics of Indigenous land tenure; and second, the impact of neo-liberal policies on the already very poor conditions in Chiapas, the Zapatistas' home state. Mexico shares with many former colonies, the wretched experiences of colonialist expropriation. The Spaniards invaded the Mexican region in the early 1500s, destroying almost half of the mostly Mayan populations that inhabited what is now Chiapas. A 1712 rebellion by Indigenous Mayan Tzelatas was brutally repressed by the Spaniards. Over a century later, Chiapas' Mayans joined the emerging Mexican state, unleashing the struggle for land that would characterize much of Mexico's future. Intense struggles between the Indigenous populations and the growing numbers of local elites and wealthy landholders continued throughout the 19th and 20th centuries. The Diaz dictatorship of 1876 to 1910, oversaw one of the most intense periods of land appropriation by local *mestizos* (landholders), as well as the opening up of Chiapas to international trade, especially coffee, cacao and timber (Hansen & Civil 2001, 445). It is no surprise that the rallying call of the ensuing Mexican Revolution was 'land and liberty'. Indigenous peasants simply sought a sustainable livelihood from land that had been theirs for centuries.

The paradox of Indigenous poverty in Chiapas is the region's wealth. Chiapas occupies only a small portion of Mexico's southeast and supports what is for Mexico a small population density of only three and a half million. It is nonetheless a very resource-rich region that contributes significantly to Mexico's export dollars. It possesses large oil and gas reserves, produces nearly half of Mexico's hydroelectric power, is the country's largest coffee exporter, and is one of Mexico's largest producer of corn, tobacco, beef, bananas, soy, cacao and timber (Burbach 2001, 117–18). The standard of living statistics tell a very different story, however. With some of the largest Indigenous populations in Mexico, particularly in areas in or around the Lacandona forest, Chiapas is the poorest and most marginalized Mexican state. In 1994, over 75 per cent of the population were malnourished, half lived in very sub-standard housing with only dirt floors, and nearly 60 per cent earned either no income or less than the minimum wage of three (US) dollars a day (see Cuninghame & Ballesteros Corona 1998; Burbach 2001, 118; Gutmann 2002). Little improved over the next decade and poverty remains critical. Infant mortality rates are high, educational opportunities for the area's Indigenous young limited, and access to medical and hospital care dire. While Indigenous poverty is widespread in Mexico, Chiapas is often considered the poorest of the poorest. In short, the experience of poverty in Chiapas is made starker by the fact of a region rich in resources and exports. The region remains semi-feudal in its concentration of productive resources in the hands of a small elite of wealthy farmers, and at the expense of the majority of the Indigenous population (Weis 2003, 14). Hence Chiapas' identity as an 'internal colony':

> The Mexican state, responding to the interests of the country's emergent bourgeoisie and the demands of the international market place, has treated Chiapas as an internal colony, sucking out its wealth while leaving its people – particularly the overwhelming majority who live off the land – more impoverished than ever (Burbach 2001, 118).

The Indigenous communities of Chiapas had long experienced marginalization and resource appropriation – well after the Mexican Revolution of 1910 which promised land and resource justice. But Chiapas did derive some post-revolution land reform. During the Mexican Revolution, revolutionaries headed by Zapata were instrumental in instigating a land reform agenda. The creation of Article 27 of the Mexican Constitution generated some land rights for Indigenous Mexicans, establishing 'inalienable and imprescriptible' ownership of land and communal holding known as *ejidos* (Blackman 2005, 106). This saw the establishment of many *ejidos*, or community farms, in the 1930s and 1940s. These farms were carved out of the previous *latifundia* land holdings controlled by the pre-revolution elites that had controlled much of Mexico for centuries. The most fertile and productive lands, however, remained in the hands of a wealthy farming elite and emerging '*nouveau riche*' ranchers. These were often appropriated with the assistance of Institutional Revolutionary Party (PRI) aligned lobbyists or *caciques*. These *caciques* 'greased the wheels of a system that still favoured the rich and well-to-do, be they in Mexico City, San Cristobal de las Casas, or in the provincial town dominated by ranchers and businessmen' (Burbach 2001, 121).

Chiapasians were nonetheless able to eke out a meagre existence through a traditional agricultural economy made possible by access to some collective lands. This traditional economy was then crippled by the series of economic shocks that convulsed Mexico. In response to the

Mexican debt crisis of 1982, where the government conceded it could no longer service its foreign debt, President Salinas introduced structural adjustment policies to take effect from the late 1980s. These policies curtailed collective land distribution and severely undermined the capacity of Indigenous Chiapasians to sustain even their meagre livelihoods. The structural adjustment packages negotiated with the International Monetary Fund (IMF) in 1982 and 1986 saw Mexico deregulate foreign direct investment policies and float its currency. Its currency underwent repeated devaluations, leading to periods of intense volatility and economic downturn. The 1994 devaluation prompted economic crisis. Amid a series of corruption scandals that dogged the 71 year-old PRI, the incoming presidency of Ernesto Zedillo Ponce de Leon negotiated a multi-million dollar IMF and US bailout. This required the introduction of major austerity measures in an effort to transform the Mexican economy into a neo-liberal one. As a result both the working and middle classes suffered under the weight of accelerating currency devaluations and rising prices for food, electricity and other staples. Unemployment rose and disillusionment with an increasingly scandal-ridden political system grew. Government opposition arose from many quarters. This included the Zapatistas – abject victims of the structural adjustment policies of neo-liberal globalization.

In this environment, momentum for land reform had stalled. As Marcos himself notes, a 'detonating' factor in politicizing the Zapatistas was the dilution of Article 27 in 1992, triggering the privatization of the *ejidos*. The land reform won by Indigenous Indians since the Mexican revolution was certainly heavily circumscribed. But the dilution of Article 27 spelt its death knell. With it died any residual hope for meaningful land autonomy for Chiapas' poorest. As Blackman (2005, 106) contends:

> The Mayan farmers were already being pushed higher and higher into the mountains and deeper into the jungles as the big ranchers moved into the more fertile plains, and they knew that with the removal of any last legal protections, as well as the flood of cheap agricultural imports that would result from NAFTA, their way of life was under serious threat.

For many Indians this tightening of their already limited land rights reinforced a history replete with domination and exploitation (see Rich 1997). This was well before the 'last straw' of structural adjustment. The collapse of coffee prices and the removal of subsidies not only on coffee but also on corn and basic grain production in the late 1980s and early 1990s, spelt disaster for local Chiapesians already struggling to sustain themselves (see Baker 2003, 294).

It seemed that something had to be done. This something came in the form of the Zapatista insurrection in 1994. The insurrection's objective was to alert the Mexican population to the dire plight of their Indigenous brothers and sisters, and to challenge the government on their behalf. While the insurrection was quickly contained, and the Zapatistas retreated to the safety of the jungles, they refused to be silenced. Unable to prevent NAFTA's implementation, the popular support generated by the EZLN uprising nonetheless worried those political and corporate sectors intent on a smooth neo-liberal transition. The Mexican army intensified their assaults on the Zapatista communities, driving them further into the jungles. Public pressure and large demonstrations in Mexico City, however, forced the government to rein back its forces and declare a ceasefire. Not only Mexico but the whole world was now watching. To pursue state repression was to ensure the Zapatistas national and global martyrdom. In any case, Zapatista appeal surged rather than waned. The dramatic unmasking of Marcos as a middle class, non-Indigenous university professor in 1995 backfired when demonstrators took to the streets chanting 'we are all Marcos!' (in Ross 2003, 13).

Increasing national and global attention on the plight of the Zapatistas compelled the Mexican government to negotiate policy compromises. The *San Andres Accords on Indigenous Rights and Culture* agreed to in 1996 promised stronger land rights and increased political and cultural autonomy (Weis 2003, 15). Any ensuing goodwill dissolved when it became clear that the Mexican government had no intention of implementing the agreement. This led to a renewal of conflict, more Indigenous deaths, increasing arrests and a renewed military presence in Chiapas. A peace agreement negotiated by a multi-party legislative body – the Commission for Peace and Reconciliation in Chiapas – failed to stem the conflict. By 1997 the Zedillo government signalled that it had abandoned the Accords and would refuse further negotiations (Harvey 2001, 1048). While shattered by these developments, the Zapatistas remained circumspect:

> We have a very important weapon which the government does not have. That weapon is called dignity. With this weapon ... nothing can defeat us. They can kill us or jail us, but they will never defeat us (Weis 2003, 15).

Hope was renewed in 2000 when the long-standing PRI lost government to the centre-right National Action Party (PAN). Incoming President Fox proved no more enthusiastic for land reform, despite promising to solve the 'Indian problem' in 'fifteen minutes'. While his parliament did finally approve a revised Accord in 2001, it was an increasingly hollow one (Harvey 2001, 1048). The Zapatistas protested and the army renewed its vigilance. A disappointed EZLN then focused more attention on improvements to their autonomous, self-governing communities. Through the creation of 'Good Government Juntas', they further democratized the five autonomous regions they controlled in Chiapas. Tensions between state officials and the communities remained even as the Zapatistas looked inward to their communities. The Zapatistas nonetheless maintained their distance and their silence. Mexico's determination to present itself to the world as a robust democracy that did not capitulate to the demands of armed minorities, contributed to the ultimate rejection of the Indigenous Rights Accord. According to Lozano (2005), had the Mexican authorities taken a more conciliatory route, 'the Indigenous communities of Chiapas could have essentially moved from armed resistance to peaceful political dialogue'; instead, they chose the traditional route of 'marginalizing the country's Indigenous communities'. But rather than containing, this forced the Zapatistas hand.

After several years of relative quiet the Zapatistas once again burst into the public domain. Issuing a 'red alert' in June 2005, they announced that they would be reconvening their communities to consider a new political direction. The world awaited their Sixth Declaration of the Selva Lacandona in which the General Command of the EZLN would detail their future course. Published in *La Jornada* newspaper over the course of a week in late June, the Sixth Declaration outlined two major changes in focus. First, they declared their intention to embrace a broader constituency of the 'dispossessed' including workers, students, Indigenous, women and a range of other marginalized groups. Their intent was to build a national and international political movement of the left to oppose neo-liberal globalization. Second, they announced their more direct engagement with the election campaign – or perhaps more accurately, anti-election campaign – in the lead up to the Mexican presidential election in July 2006. Planning for the *La Otra Campana* – the Other Campaign – began in August 2005 with a series of participatory meetings between the Zapatistas and a diverse range of constituencies. It was agreed that, from early 2006, a 'travelling circus' made up of Marcos and a small group of other Zapatista leaders would make their way across Mexico, talking and listening to many grassroots and workers organizations. They aimed to gather impetus for a broad left coalition of the disaffected. In short, they were on a 'drive to consolidate the non-electoral anti-capitalist left' as a whole (Ross 2005). The Other Campaign is an ambitious, far sighted project, expected to continue after the July election into 2007 and beyond. But unlike their political party counterparts, the Zapatistas 'have a longer-range political goal than taking power'; they want to organize Mexico for a new constitution and build the momentum for change from the grassroots and the bottom up (Ross 2005).

While there has been much speculation in the media that the Zapatistas' new direction might involve traditional party politics, or even the formation of a new left party, their blanket condemnation of all political parties makes this highly unlikely. Referring to the hollow Accord of 2001, Marcos declared that all major parties had let them down. These parties 'had their chance' but 'did not keep their word'; it was now quite clear that 'there was no point to dialogue and negotiation with the bad governments of Mexico' (Marcos 2005, 3). Marcos reiterated that all political parties of whatever hue provided only 'tricks, lies, mockery and disdain' (Wissert 2005). Nonetheless, it seems that their national tour could influence the outcome of the 2006 Mexican elections, even if Marcos refers to himself as 'delegate zero'. Despite much talk of their demise and increasing irrelevance, the Zapatistas' re-emergence captured widespread attention, both nationally and globally. It was clear that they were not yet a spent force.

The Zapatista National Liberation Army

The Zapatistas' dramatic entrance onto the Mexican and global stage posed the question of who they were and where they came from (Ross 1995). Their origins begin with the National Liberal Forces (FLN), formed after the student massacres of 1968 that saw hundreds of student demonstrators killed on orders from the PRI-led government. Informed by a combination of Marxist, Maoist and Che Guevarian politics, the FLN quickly went underground as the government sought to contain its spread. Members of the FLN, which included a young Marcos, then formed the EZLN – *Ejercito Zapatista de Liberacion Nacional* or Zapatista National Liberation Army – in 1983. The EZLN was invited into the first Indigenous community in 1986, and grew to over 1300 armed members three years later (Hansen & Civil 2001, 447). While its principal goal was 'land and liberty' for Chiapas' Indigenous, the focus gradually broadened to incorporate all those marginalized groups that shared the experience of repression and dispossession. Its expanded focus prompted a change in its 'ideology', especially after the 1994 insurrection. In particular, it cast off any resemblance of a Marxist-Leninist revolutionary vanguard. Marcos' articulation of the Zapatista political philosophy was critical to the EZLN's renewal. Addressing the National Indigenous Forum in January 1996, he makes clear that autonomy and diversity are now central:

> Each one has his own field, his own planting, but we all have the same village, although sometimes we speak different languages and wear different clothes. We invite each of you to plant your own plot and in your own way. We invite you to make of this forum a good tiller and makes sure that everyone has seed and that the earth be well prepared (Marcos 1996b, 85).

Reiterating this message during the global Eucentro gathering in Chiapas in August 1996, Marcos declared:

> This intercontinental network of resistance, recognising differences and acknowledging similarities, will strive to find itself in other resistances around the world. This intercontinental network of resistance will be the medium in which distinct resistances may support one another. This intercontinental network of resistance is not an organising structure; it has no central head or decision maker; it has no central command or hierarchies. We are the network, all of us who resist (1996c, 117).

This change in direction contrasts with the politics of the First Declaration. Invoking the discourse of class struggle, the 1994 uprising quickly cast the EZLN, even if erroneously, as a quasi-Marxist guerrilla movement. Through their declaration of war, the First Declaration posed armed liberation and insurgency as the preferred route of political change. Invoking elements of vanguardism and nationalism to 'restore the legitimacy and the stability of the nation by overthrowing the dictator'; and addressing its 'Mexican brothers and sisters', the EZLN's First Declaration asserted:

> We are a product of five hundred years of struggle: first, led by insurgents against slavery during the War of Independence with Spain; then to avoid being absorbed by North American imperial-

ism; then to promulgate our constitution and expel the French empire from our soil; [and] later when the people rebelled against Porfirio Diaz's dictatorship, which denied us the just application of the reform laws, and leaders like Villa and Zapata emerged, poor men just like us (Marcos 1994b, 13).

In the Second Declaration of June 1994, they announced that they would 'not surrender', even as they committed to the negotiated cease-fire. Importantly, they also declared that 'democratic change is the only alternative to war'. At this stage they still entertained institutional reform and focused on the upcoming election, contending that a new kind of grassroots political party 'must be born'. These institutional appeals gradually weakened, not least because of the ongoing disappointments of the land reform process.

A less militaristic and revolutionary identity contributed to their growing national and global appeal. This served strategic as well as political objectives. As strategy, global awareness of the EZLN presence ensured them protection within their own borders. Mexican governments may have been eager to quash the movement – their firepower was after all far superior – but they risked their political reputation in the face of an attentive world. Politically, the Zapatistas extended their appeal through their increasing dissociation from a Marxist revolutionary left whose reputation had been tarnished by some of its authoritarian forays. Their stronger embrace of democracy, diversity and autonomy helped broaden their appeal considerably – even as it infuriated the old left. But the Zapatistas did not throw off all 'traditional' influences. Their continued allegiance to Zapata needs special mention.

From Zapata to Zapatista

Emiliano Zapata is the Zapatistas inspiration, as reflected in their naming. A Nahua Indian, Zapata was an important figure in the Mexican revolution of 1910. He was one of the first to crystallize fighting for Indigenous Indians land rights as key to their future livelihoods. Zapata's resistance grew out of the growing working class struggles against an encroaching industrialization built on colonialist appropriation of land and exploitation of labour. Against this background, the Marxist and anarchist call to arms resonated in a continent that had already established strong socialist and anarchist roots. Anarchist ideas were brought by the European immigrants, especially Spanish and Italians, who flooded South American shores during the second half of the nineteenth century. The continent's untapped resources attracted many entrepreneurial industrialists, but with them also came growing numbers of labourers and peasants that opposed them. Anarcho-syndicalism, Zapata's radical 'brand', was particularly influential. It dominated the South America working class movement until at least the 1930s, with many large-scale struggles between the state and anarchists assuming 'undeclared civil war' proportions (Marshall 1993, 504). Links between levels of industrialization and anarchism, saw some countries develop much stronger anarchist movements than others (Marshall 1993, 504–5). Argentina, Brazil and Uruguay had strong anarchist movements, but it was in Mexico that anarchists directly participated in revolution.

Prior to the seizure of large parcels by wealthy landlords, land in Mexican villages had been held, worked and shared communally among peasants and Indigenous Indians. With the support of the Diaz dictatorship from the late 1880s on, increasingly larger parcels of land were appropriated from the communities and into private hands (see Katz 1996). This fomented growing resistance and guerrilla activity. Impassioned revolt was led by the anarchist Flores Magon brothers whose resistance paved the way for the Mexican Revolution. Witnessing the pillage of many villages and the brutal appropriation of their lands, Zapata had also been active in forging resistance against the invading landlords, especially in his own home state. During the revolution itself, Zapata led the peasants and Indians in the reclamation of their stolen lands. As Marshall (1993, 511) notes, while 'primarily an egalitarian movement which sought the redistribution of the land and the right to be left alone' these rebels increasingly 'resembled the peasant anarchists ... of the Spanish Civil War in their moral purity and contempt for politics'.

After a successful coup against Diaz, the more moderate and reformist Madero formed government. By now entirely distrustful of social transformation through political channels – an approach he had at first entertained – Zapata led an ultimately unsuccessful rebellion against Madero. It was Zapata himself who was betrayed and killed in 1919. It seemed that with his death, and that of Flores Magon in exile, also died any genuine hope that the revolution would transform the fabric of Mexican society into one based genuinely on liberty and land equality. Throughout these turbulent years, it was clear that Zapata 'died as he lived, an honest and courageous peasant, fighting for land and liberty for his people' (Marshall 1993, 513). This simple moral quest lives on in the Zapatista spirit. Zapata was not a committed and ideological anarchist in the same way as Flores Magon. In fact Zapata's ideological identity is still contested, with Marxists, socialists and anarchists alike claiming him as one of their own. But his devotion to land and liberty, his suspicion of political reform and centralized power, and his commitment to community autonomy and Indigenous dignity, helps explain why both Zapatistas, and anarchists, invoke the spirit of Zapata in their struggle against dispossession (see Esteva 2001).

On the April anniversary of Zapata's assassination, Marcos combined the spirit of both Zapata and the Indigenous Tzeltales in his naming of Votan Zapata as the movement's emblem:

> Brothers and sisters, we want you to know who is behind us, who guides us, who walks in our shoes, who rules our heart, who rides our words, who lives in our deaths ... Votan Zapata ... the name that changes, the man without a face ... looked in Miguel, walked in Jose Maria, was Vicente, named in Benito, flew in a bird, rode in Emiliano, shouted in San Francisco, visited in Pedro ... Votan Zapata, guardian and heart of the people. And in our nameless, faceless path, he took a name in us: the Zapatista National Liberation Army ... Zapata, in being arrives! In death he lives! Viva Zapata! (Marcos 1994a, 19–21).

As the first human emissary sent by God to distribute land among the Tzeltales, the early Votan clearly shared the same goals as the later Zapata. Both are invoked to live on through the Zapatista spirit.

EZLN organization and structure

The Zapatista movement is compromised of three elements: the armed wing – the EZLN; the Zapatista communities in the jungles of Chiapas; and 'Frente Zapatista', the movement's nationwide support organization. Overseeing all these elements is the Clandestine Indigenous Revolutionary Committee headed by a small number of Subcommandantes, of which Marcos is a key figure. Information on the EZLN's organizational structure is necessarily patchy. Understandably cautious, the organization and the communities guard against intrusion for fear of political and military reprisal. Security is hence paramount. Most information comes from Marcos and the EZLN committees themselves, as well as from the carefully screened sympathizers who are permitted into their midst for short periods of time. Nonetheless, a glimpse into its organizational culture is important for establishing both its democratic and anarchical character. First and foremost, the

EZLN is an army and as such adopts a militaristic structure that is hierarchical and disciplined. As Marcos himself reminds us, the EZLN is a clandestine organization that heads a movement that resists and rebels against a formidable, and well-armed, foe. At the same time, Marcos claims that the EZLN was originally conceived as a force for self defence (in Benjamin 1995, 65). The decision to go to war was born from this need to defend their communities from encroaching ranchers who had the blessing of the Mexican state.

The EZLN is nonetheless an unusual 'army'. We do know that EZLN leaders cannot own property or hold political office: 'Unlike the tradition of "heroic" guerrilla commanders', they have made efforts to flatten military hierarchies through single 'leadership' layers – they are, for example, all commandants, and often mock their own 'leadership' pretensions (Burbach 2001, 117). But the EZLN is still an army, and armies contain hierarchies and chains of commands that are inherently undemocratic. While the EZLN has made democratic changes, its internal organization retains undemocratic features. In a 2001 interview, Marcos had no hesitation in proclaiming that 'we were formed in an army, the EZLN. It has a military structure [and] Subcommandante Marcos is the military chief of an army'; but he also points out that 'our army is very different from others, because its proposal is to cease being an army' (Marcos in Marquez & Pombo 2001, 70). Its internal structure may not be democratic but, importantly, its overall relationship to the communities is. Significantly, their decision making is guided by consultations with the community rather than emanating directly from EZLN's 'hierarchy'.

According to Marcos, the decision to go to war in 1994 was made by the communities who 'told me to start the war, because I was in charge of military planning' (in Benjamin 1995, 65). When Marcos argued that they were not ready to go to war in 1993 as directed, he was given one more year 'to make the arrangements', and otherwise told that if he did not do it in a year, it would be done without him (in Benjamin 1995, 65). While most of the final strategic details derive from this Committee, Marcos emphasizes the democratic logic of its structure:

> I cannot command militarily. That is what no one understands. Marcos does not need the Clandestine Committee as a justification, because that is undeniable. In order to make political decisions and to be able to exert the military command, Marcos needs the authority that prevails over communities. He needs the communities per-

mission to give the order of war, in order to give the order of withdrawal. If I don't have that permission, I do not exist as a military commander ... and the EZLN does not exist ... The moment the communities say: 'You go' ... I gave to go. Or I risk being left alone (in Blixen & Fazio 1995).

For Flood, 'unlike almost all other rebel armies, the command of the army does not end in its own military command but rather in the hands of those at the base whom it claims to represent' (2001, 11). One of the reasons for this lies in the EZLN's very evolution. It grew from a few students scaling the mountains of Chiapas intent on leading a people's liberation effort, to one that 'was forced to accept that the people and not the army command should have the final say' (Flood 2001, 11).

The system of delegate democracy is central to the Zapatistas 'home grown' democracy. In making all levels of the EZLN answerable to the ordinary community members at the base of the structure, the communities are able to enact a decision-making network that incorporates the views of thousands and that enables a decentralized structure of communities to function as a collective whole. In this way, the EZLN becomes the Zapatistas' implementing body. Like most forms of delegate democracy there are mechanisms in place to restrict delegate power. These include limiting their tenure and their capacity to speak for the organization, while strengthening consultation between the organization's different levels. Referring to its 'consultative assemblies', the EZLN claims that:

It is the communities who elect their representatives for the Autonomous Municipality Council which is the authority for the municipality. Each representative is chosen from one area of administration from within the autonomous municipality, and they may be removed if they do not comply with the communities mandates ... [T]hose who hold a position on the municipal council do not receive a salary for it, although their expenses should be paid by the same communities who request their presence (in Watson 2002, 77).

Even so, the Zapatista communities do not claim to have perfected their internal democracy and continue to struggle with exclusionary practices. Particularly problematic are the role of women and traditional Indigenous structures that privilege the authoritative role of elders. In the 'Two Flaws' communiqués of August 2004, Marcos acknowledges these problems and sets about redressing them. These communiqués focus specifically on tackling the 'mistakes which seem to have persisted in Zapatista political work: the place of women, on the one hand, and, on the other, the relationship between the political-military structure and the autonomous governments' (Marcos 2004). These communiqués offer insights into the size of the democratic task confronting the communities:

> It will take time, I know. But for those who, like the Zapatistas, make plans for decades, a few years isn't much time ... The fact that the EZLN is a political-military and clandestine organisation still corrupts processes that should and must be democratic ... These two flaws need our special attention, and obviously, measures to counter them. We cannot blame the military encirclement, the resistance, the enemy, neoliberalism, the political parties, the media, or the bad mood that tends to accompany us in the mornings when the skin we desire isn't there (Marcos 2004).

These problems had been recognized for quite some time. Writing in 2001, for example, Comandanta Ester states:

> We know which are good and which are bad *usos y custombres*. The bad ones are hitting and beating a woman, buying and selling, marrying her by force against her will, not being allowed to participate in assembly ... [yet we struggle] to be included in this law [San Andreas Accords], so that no one will any longer be able to prevent our participation, our dignity and safety in any kind of work, the same as men (in Gallaher & Froehling 2002, 94).

The Juntas of Good Government formed in 2004 were tasked with the express purpose of addressing these problems and the further democratization of the communities as a whole. The region's extreme poverty, low literacy and poor infrastructure, complicates the realization of these democratic objectives. It also highlights the foresight required to attempt them.

Critics nonetheless charge that the Zapatista democracy is all smoke and mirrors, and they are intent on subverting rather than promoting the democratic process. A common observation is that what the mask truly hides is the fact of an authoritarian organization intent on violent resistance. Marcos himself is often portrayed as an anti-democratic figure set on capturing personal glory. His mask hides his identity as an urban intellectual rather than a poor Indian, and the fact that he has retarded rather than helped the Indigenous poor for whom he seemingly advocates. In light of this, some doubt that 'any reputable left-winger will take up his offer' to establish a new coalition of the left (R. Delarbre Trejo 2005, pers comm. 28th June).

Zapatismo

Zapatismo is the political philosophy of the Zapatista movement. It is both a vision and a strategy for social change. Its overall aim is to 'spark a broadly-based movement of civil society in Chiapas and the rest of Mexico that will transform the country from the bottom up' (Burbach 2001, 116). Through restoring them their dignity, it represents the tenacity and hope of the oppressed. Zapatismo demonstrates that history cannot be swept aside and the refusal to address past wrongs will always reanimate resistance. While wanting to change the world, Zapatismo does not seek to *take* power, but to *exercise* it. Decidedly of the left, Zapatismo is committed to the principles of just distribution, human dignity and freedom. It is not beholden to any ideology, and hence answers to no 'position'. Rather, it champions its *absence* of ideology. Marcos describes it this way:

> Zapatismo is not an ideology; it is not a bought and paid for doctrine. It is ... an intuition. Something so open and flexible that it really occurs in all places. Zapatismo poses the question: 'What is it that has excluded me? What is it that has isolated me? In each place the response is different. Zapatismo simply states the question and stipulates that the response is plural, that the response is inclusive (in Couch 2001, 244).

The struggle for autonomy sits at the heart of the Zapatista struggle (see Esteva 2001). It will not describe in any detail the new world it seeks to make. To do so is to take its envisioning out of the hands of the autonomous communities to which it belongs. It does not represent *a* singular way of changing the world. Rather it resurrects the possibility of change and passes the form and process of its making over to its envisioners. It is also here that Zapatismo invokes considerable Third World and global solidarity (see Olesen 2004).

For Marcos, embracing the *possibility* of change is in itself a radical act. It undermines the political paralysis and hopelessness that neo-liberalism has imposed on the oppressed worldwide. Addressing the First Intercontinental Eucentro in 1996, he states: 'A new lie is sold to us as history. The lie about the defeat of hope, the lie about the defeat of dignity, the lie about the defeat of humanity' (in Couch 2001, 243). Zapatismo is thus as much a 'state of mind' as a political strategy, with hope for social change resurrected in two main ways. First, it proposes that even when all possibilities seem exhausted, there are ways of breaking through. The 1994 insurrection represented enormous hope against enormous odds. It was its brazenness and courage that resonated worldwide. Second, Zapatismo endorses a plurality of struggles and a plurality of actions to replace the sectarianism that demanded a stifling conformity and a strategic singularity. Writing in 1994, Marcos asserts that:

> Something broke in this year, not just the false image of modernity sold to us by neo-liberalism ... but also the rigid schemes of the Left living in and from the past. In the midst of this navigating from pain to hope, political struggle finds itself naked, bereft of the rusty garb inherited from pain: it is hope which obliges it to look for new forms of struggle, that is, new ways of being political, of doing politics: a new politics, a new political morality, a new political ethic is not just a wish, it is the only way to go forward, to jump to the other side (in Couch 2001, 246).

It is this 'new politics' that animates Zapatismo.

In his closing remarks to the 1996 Encuentro and in what many have claimed comes as close to a 'manifesto' of Zapatismo as is ever likely, Marcos states:

> On the one side is neoliberalism, with all its repressive power and all its machinery of death; on the other side is the human being. There are those who resign themselves to being one more number in the huge exchange of Power. There are those who resign themselves to being slaves ... [to] the crumbs that Power hands out ... But there are those who do not resign themselves ... those who resist ... those who decide to fight ... In any place in the world, anytime, any man or any woman rebels to the point of tearing of the clothes resignation has woven for them and cynicism has dyed gray. Any man or woman, of whatever colour, in whatever tongue, speaks and says to himself or to herself: Enough is Enough. – *Ya Basta!* ... Let [this echo] be a network of voices that resist the war Power wages on

them ... the multiplication of resistances ... the many worlds that the world needs ... (Marcos 1996d, 110–15)

Mexican author, intellectual and Zapatista sympathizer, Gustavo Esteva captures well the uniqueness of Zapatismo. He observed that the Zapatistas were distinctive from their very outset. They were disillusioned with the traditional route to justice such as the ballot box, and tired of the 'give us your state and we will do it better' type of socialist promise (in Kingsnorth 2003, 43). Instead they propose 'alternatives to both', a new kind of radical democratic politics. Esteva contends that:

> Here is a new way of looking at the world. Take the Zapatistas. They call thousands of people down here to Mexico for the Encuentro, and when they get here the EZLN say 'don't follow us, we will not be your vanguard'. Why? Because they say they don't have the truth, and they should not lead anyone else. Perhaps they have a truth – a truth for the Chiapas, but not a universal truth, that can apply everywhere. All over the world, there are other truths. In other places, perhaps the principles can be applied – radical democracy, at the grass roots, claimed by people who are linked together worldwide. But the way it manifests itself can be different everywhere' (in Kingsnorth 2003, 43).

Zapatismo is nonetheless shaped by its history and it culture. This includes its nationalist roots and its history of Indigenous struggles. Some look no further than its historical roots when conceptualizing the Zapatista movement as essentially a nationalist movement, one that utilizes the discourse of nationalism to further its political ends (Gallaher & Froehling 2002). Yet the discourse of nationalism resonates in a particular way for Mexico's Indigenous. Their usually poorer and more marginalized position in Mexican society made a mockery of historical calls to see themselves as Mexicans rather than Indigenous. Seldom enjoying the same privileges, and hence seldom fully Mexican, the Indigenous turn towards autonomy could not have been entirely unexpected. But not all Indians and peasants in Chiapas support the Zapatistas and many are highly critical. They particularly denounce what they interpret as a push for separatism – a push the Zapatistas deny. The Zapatistas reiterate that autonomy is not the same as separatism. Commandante Susana of the Clandestine Indigenous Revolutionary Committee (in Gallaher & Froehling 2002, 92) explains: 'It's not true that we want to separate from Mexico'; instead we want 'them to recognise us as

Mexicans, and the Indigenous we are, but also as Mexicans, since we were born here, we live here'. In short, the Zapatistas of Chiapas wish to be included as full – but autonomous – members of the Mexican community. In the language of Zapatismo, the achievement of such goals requires the exercise rather than taking of power. It is the discourse of power that distinguishes Zapatismo. It is also where Zapatismo intersects most robustly with anarchism.

Zapatismo and the anarchist impulse

Zapatismo has strong synergies with post-ideological anarchism, even as it draws from a diverse ideological constellation including Marxism, socialism, Maoism, liberation theology and anarchism. Some of the EZLN's early recruits originated from Maoist groups formed in the 1970s, and Marxism has had a long association with guerrilla movements in developing countries. Liberation theology has been particularly influential in Latin America. While opposing armed struggle, the popular Catholic Bishop from San Cristobal, Samuel Ruiz, helped educate and mobilize the peasants, organizing them into cadres and unions that could challenge the Mexican state (Castells 2004, 78). Despite these influences, the Zapatistas refuse *an* ideological identity. They are clearly not intent on overthrowing the established order. Nor do they seek to replace a government they intend to conquer. Instead, they seek the very opposite: an absence of state power. But they seek the *exercise* of power where it matters most: to achieve autonomy for their communities, sufficient resources to sustain themselves, and a democratic culture underpinned by inclusive politics. Marcos states:

> We are undefinable, we don't fit in any category: it is not clear whether we are Marxists, if we are anarchists, if we are neither one nor the other, if we are indigenists ... nationalists or internationalists. The Movement is that undefined that it can fit any models ... The moment that Zapatismo acquires a theoretical body or an organic level, it will disappear or it will finish like many doctrines that we have seen vanish on the stage (in Aviles & Mina 1998, 176).

In a 1995 communiqué, Marcos complains that for communists he is too much of an anarchist; for anarchists he is either too much of a communist; for radicals he is too reformist and for reformists he is too radical; the proletariat accuse him of excluding them and civil society charges he expects too much of them. But it is this refusal to be pinned down ideologically that distinguishes Zapatismo.

While the Zapatistas do not classify themselves as anarchist, and reject the whole process of political classification, they draw on a number of key anarchist principles. For some observers their tactics are 'perfectly in accord with the general anarchist inspiration of the movement, which is less about seizing state power than about exposing, delegitimizing and dismantling mechanisms of rule while winning ever-larger spaces of autonomy from it' (Graeber 2002, 68). In defending their refusal of 'exemplary' socialism – a defence that would have heartened anarchists – Marcos proclaims:

> We are not a proletariat, our land is not your means of production and we don't want to work in a tractor factory. All we want is to be listened to, and for you big-city smart-arses to stop telling us how to live. As for your dialectic – you can keep it. You never know when it might come in handy (in Kingsnorth 2003, 29).

But while rejecting ideological socialism, and vanguardism in particular, Zapatismo has not been averse to utilizing the language of class struggle to rouse support for their cause (see Couch 2001, 250). Marcos does not deny his Marxist past, nor does he reject Marxism altogether. He is simply intent on refusing the stranglehold of ideology and denying any ties or favours to Marxists, or others. The Zapatista 'revolution', unlike its socialist counterparts, 'will not end in a new class, faction of a class or group in power' but rather 'in a form of democratic spaces for political struggle' (in Watson 2002, 75). As Holloway points out, Zapatismo goes beyond 'the state illusion' that placed the state at the centre of radical change and that identified revolution 'with control of the state' (2002b, 157).

While they reject the revolutionary left's capture of the state, the Zapatistas nonetheless negotiate a more flexible relationship to it than ideological anarchists would condone. They seek neither to eradicate it nor a total separation from it. Reflecting a nationalist core they share with many other guerrilla movements, they seek instead to 'reconstruct' the idea of a nation. But this reconstructed nation would be 'rebuilt on different foundations ... based in the recognition of difference' (Marcos in Gallaher & Froehling 2002, 93). The recognition of difference reconceptualizes the state to privilege civil society. This 'privilege' is not conceived in vanguardist terms, however. Rather, it is discursive – establishing words and not guns as the new weapons wielded by civil society actors. Facets of this discursive politics are informed by a Gramscian analysis. The Zapatista refusal to conquer state power, unlike many guerrilla movements, resonates Gramscian insights. As Johnston (2000, 467) notes, Gramsci's 'conceptualization of a war of position is a useful tool for understanding an armed struggle that target ideas'. When total revolution, or 'war of movement', is not possible, a powerful alternative is 'a war of position that targets ideas, attitudes, the state, and civil society'; its aim is to build 'a broad counter-hegemony, while resisting co-optation by more powerful hegemonic forces' (2000, 467).

Civil society helps drive this counter-hegemony. Against a state corrupted by the excesses of political power, Zapatismo pits civil society:

> [Civil society is] organised itself little by little to become proof that you can participate without aspiring to public office, that you can organise politically without being in a political party, that you can keep an eye on the government and pressure it to 'lead by obeying', that you can have an effect and remain yourself ... the strength of Civil Society that so perturbs government leaders, today gives us hope that it's possible to rebuild the country despite the destruction the neoliberal project has brought to the Mexican society (Marcos 1996a, 121).

The two forces of civil society and a corrupted Mexican state confront each other: 'On the one hand there is their nation, their country, their Mexico. A plan for the nation that Power holds up with bloody hands, with law and legitimacy soiled by corruption and crime ... the Mexico that belongs to Power'; on the other, civil society: 'the only force that can save the country' (Marcos 1996a, 122).

Importantly, it is in what the Zapatistas oppose to the state, or what they intend to replace it with, that distinguishes their relationship to it. They have little time for a hierarchical state – socialist or otherwise. Rather, the autonomous, municipal and delegate structure of the Zapatista communities presents 'a clear alternative to the state'; and it is this process that 'truly determines the health of a revolution rather than the fine words of its leaders or the slogans it is organized under' (Flood 2001, 15). The opposing of inclusive decision making from below against an exclusive centralized state represents its radical remaking. The Zapatistas intend that 'groups in power will be watched by the people in such a way that they will be obligated to give a regular account of themselves and the people will be able to decide whether they remain in power or not' (EZLN in Watson 2002, 75). As Baker (2003, 303) points out, the Zapatistas aspire to creating 'alternative *counter*-publics, as if in recognition that there will always be a dominant (statist) public that practises various forms of exclusion'. The Zapatistas' anti-statism consists not of direct attempts to abolish it, but of ensuring autonomous spaces for its citizens, so that they can 'organize politically without becoming a political party', so that they can keep the government accountable and pressure it to 'lead by obeying', and so they can have an effect but retain their identity (Marcos 1996a, 122). This civil force that exercises rather than takes power becomes 'the only force' that can save Mexico.

Zapatismo and power

It is Zapatismo's conceptualization of power that resonates anarchist sensibilities most acutely. The Zapatistas want to *exercise* power (that is direct their own lives autonomously and thus dissolve the relations of power) rather than *take* it (assume centralized, authoritarian and statist reins of power). In this way they challenge the very notion of power. They grapple not with *'whose* power but the very existence of power' so that what is 'at issue is not *who* exercises power, but how to create a world based on mutual recognition of human dignity, on the formation of social relations which are not power relations' (Holloway 2002a, 17–18). For Marcos the purpose of social change is the creation of 'a space in which people can define their own power' thus birthing a 'new political practice' that moves beyond a crude seizure of power to the very 'organization of society' (in Flood 2001, 3–5).

Marcos thus distinguishes between a capitalized Power and a lower case power. This is an important distinction. It not only goes to the heart of Zapatismo's anarchist temperament but also clearly demarcates itself from a Marxist/Leninist vanguardism. Capitalized Power is embraced by a corrupted state that utilizes the tools of legitimacy to *take* power rather than *exercise* it. This Power imposes its hegemony *over* 'its' people and represents the 'Mexico that belongs to Power'. In its stead, Marcos proposes the democratic exercise of a lower case power that is internal rather than external to each individual member of the civil community. In a 2001 interview he states:

> There is an oppressor [Power] which decides on behalf of society from above, and a group of visionaries who decides to lead the country on the correct path and ousts the other group from power,

seizes power and then also decides on behalf of society. For us that is a struggle between hegemonies ... [where] things basically don't change ... You cannot reconstruct the world ... on the basis of a quarrel over who will impose their hegemony on society (in Marquez & Pombo 2001, 70).

In short, only a democratic lower case power can undermine the hegemony that is Power. In seeking to change society from below, Zapatismo thus acknowledges the anarchist caution against the centralization of power, a Power which always 'turns against itself'.

Zapatismo also shares anarchism's circumspection about institutional change, particularly through electoral democracy. Marcos observes that while Mexico gives its citizens the vote, 'you cannot vote for an alternative path', simply more of the same (Benjamin 1995, 58). The Zapatistas thus refuse official political positions (Marcos was even 'invited' for talks with President Fox in 2005) in the 'kingdom' of political power. Acknowledging that electoral democracy and political parties are necessary for operationalizing democracy, in a June 2000 Communique Marcos asserts that they are not sufficient conditions:

Election time is not the time for the Zapatistas ... We want to find a politics which goes from below to above, one in which 'governing obeying' is more than a slogan; one in which power is not the objective ... Concerning the political parties, we say that we do not feel represented by any of them ... We criticise the parties' distance from society, that their existence and activities are regulated only by the election calendar (in Flood 2001, 19).

These views remain consistent with the Zapatistas' nationwide tour to dialogue with those from below, in the lead up to the 2006 Mexican elections.

Against the hollowness of electoral democracy, Zapatismo proposes a radical democracy. The Zapatista democracy is inclusive and built from below. While they see some value in electoral politics 'properly conducted', this form of social organization can never replace radical democracy's legitimation of autonomy and difference. The Zapatista conception of democracy blends forms of unmediated and direct democracy (as practised within their communities) with mediated and indirect ones (as members of the Mexican state). Initially prepared to engage with the political arena for land reform, continued disappointment reinforced the Zapatistas' intuitive distrust of institutional politics. Zapatismo's radical democracy resonates elements of Laclau and Mouffe's version of it. The Zapatistas agree with them that task of the left 'is not in the abandonment of the democratic terrain but, on the contrary, in the extension of the field of democratic struggles to the whole of civil society and the state'; in short, the struggle is 'in the direction of a radical and plural democracy' (2001, 176). In his investigation of the Zapatista form of radical inclusive democracy, Watson (2002, 86) concludes that the EZLN's democratic project galvanizes a 'potent resistance' to neo-liberal globalization by urging direct participation in the generation of alternatives that '[chip] away at the edges of dominant thinking and practice'. Many others concur. Importantly, this radical democracy is open to building on any genuine democratic impulses in already existing institutions. In this way, the Zapatista struggle is 'a critical piece of a pro-democratic movement that challenges the shape of actually existing Mexican democracy' (Johnston 2000, 492). Their 'revolution' is not in the undoing but rather in the doing of democracy. It is the creation of a world in which 'all worlds are possible'.

Zapatismo commits to an inclusive pluralist democracy, unlike many past revolutionaries. Marcos identifies 'two major gaps' in Latin America's revolutionary left. The first excludes Indigenous peoples and the second denies a broad array of other 'minorities'. Rather than simply oversights, these exclusionary practices underpin the very edifice of Marxism-Leninism (Marquez & Pombo 2001, 71). Unlike other vanguards, Zapatismo speaks 'in the name' of minorities rather than majorities. To claim to represent or speak for majorities is 'beyond ourselves' and thus an illusion. But to speak 'in the name of' is not the same as leading: 'we had not come to lead anything of what might emerge. We came to release a demand that could unleash others' (Marquez & Pombo 2001, 72). In short, each minority must articulate and situate their own grievances – become, in effect, their own Marcoses:

> Marcos is gay in San Francisco, a black person in South Africa, Asian in Europe ... a Jew in Germany ... a woman alone in a Metro station at 10 pm ... a writer without books or readers, and a Zapatista in the Mexican Southeast ... [Marcos] is every minority who is now beginning to speak and every majority that must shut up and listen ... Everything that makes power and the good consciences of those in power uncomfortable – that is Marcos (in Kingsnorth 2003, 30).

The Zapatista mask is emblematic of both the shared experience of different oppressions – 'we are all Marcos'; as well as representing the 'cover up' that is the real Mexico:

> I will take off my ski mask when Mexican society takes off its own mask, the one it uses to cover up the real Mexico. Then Mexicans would see that the self-image they have been sold is false ... And once they have seen the real Mexico – as we have seen it – they will be more determined to change it (in Benjamin 1995, 71).

Zapatismo echoes the central anarchist insight that the ends can never justify the means, since the means are the ends. The Zapatistas define their goal 'by the way we choose the means for struggling for it' (Marcos in Marquez & Pombo 2001, 76). Central to this insight are their operational principles of 'command-obeying' and 'asking we walk'. First and foremost, command-obeying subverts hierarchy by juxtaposing and inverting the relationship of 'leaders' and 'led'. 'Asking we walk' positions the process of social change at the centre of 'revolution'. Revolution and liberation depends on the process of asking the right questions rather than providing the right answers. Unlike old vanguards who have always claimed to know both the questions and the answers, there is in fact no kind of 'final solution' or teleological end point. Social change is a dynamic process, the detail of which can only be arrived at in the process of doing, or asking. 'Command-obeying' and 'asking we walk' also reflects Marcos' own personal journey. Arriving in the jungles of Chiapas in the early 1980s armed with Marxist ideas with which to 'educate' and radicalize the poor, he soon realized that the people knew how they wanted to live even if they lacked the means to do so. This prompted his own 're-education', one that practised listening to rather than speaking for others, and that paved the way for the post-ideological politics of Zapatismo (Tormey 2004, 131, 129–38).

Long time observer and sympathizer, Luis Hernandez Navarro, from Mexico's *La Jornada*, also identifies the Zapatista distinctiveness in their 'non-ideological' stance (L. Hernandez Navarro 2005, pers comm. 27ᵗʰ June). By this he means that 'what they say and what they think and what they do, comes from the roots, not from ideas from outside'. Theirs is not an ideology 'from the outside' but rather 'something that became from their own experience'; but 'it's not just an idea, it's a practice, and that makes ... a big difference' (L. Hernandez Navarro 2005, pers comm. 27ᵗʰ June). Zapatismo is 'not a conceptual theory that is complete'; rather 'it's a mix, but that mix makes something new'. Importantly, he identifies the Zapatistas as a rebel rather than revolutionary force. Key to this distinction is their conceptualization of power. Whereas a revolutionary force seeks to take power, a rebel force seeks simply to exercise it. Hernandez Navarro also identifies the Zapatistas' ideological refusal, noting that they cannot be neatly categorized; for the Zapatistas theory is in any case 'just practice'. Holloway makes similar observations. He notes that the contemporary challenge to 'change the world without taking power' has been taken up most convincingly by the Zapatistas. Theirs is an 'anti-power' where the 'old distinctions between reform, revolution and anarchism no longer seem relevant, simply because the question of who controls the state is not the focus of attention' (Holloway 2002a, 21). But anti-power is not a counter-power, rather 'the dissolution of power-over' and 'the emancipation of power-to' (2002a, 36).

Conclusion

This chapter has demonstrated the Zapatistas' notable success in catapulting a local struggle onto the global stage. Their political philosophy helped create a distinctive form of resistance that transcended traditional dissent and inspired oppositional politics worldwide. Locally, the Zapatistas have carved out autonomous zones for their communities in the mountains of Chiapas. These zones are organized according to democratic, inclusive and reflexive principles. Nationally, their presence discomfits Mexican governments and can affect electoral outcomes. At times of writing, they are continuing their nationwide 'tour' to dialogue with the grassroots against the background of a looming presidential election. Globally, they have stimulated the impetus for social change and the anti-globalization movement owes them much. Despite these local, national and global successes, the fact remains that Indigenous poverty continues unabated, and their struggles have a considerable way to go. Not unexpectedly, there also remain paradoxes and contradictions in Zapatismo. The Zapatistas are both 'pedagogical guerrillas' and 'armed democrats'; they are an army that has led armed insurrection but one that also hosts democracy conferences and seeks to flatten hierarchies (Johnston 2000). And despite shunning leadership, in the conversion of a grassroots rebel into a global 'celebrity', the Zapatistas have seemingly continued the Latin American penchant for charismatic leadership.

In Marcos, the Zapatistas boast a charismatic figure and a global 'star' who captures the world's romantic as well as political imagination. Arresting images of a masked and handsome superhero – a poetic Davidian everyman battling globalization's Goliath – makes for compelling media. As the 'professional of hope' Marcos was quickly embraced as an oppositional superstar. But Marcos was no pawn in a global media's game. Acute to the impact of the media event and the iconic imagery that drove it, Marcos constructed himself as both icon and myth. Moreover, his global 'stardom' protected the EZLN against a retaliatory Mexican state. Neither has living in the bowels of the Mexican jungles precluded this 'postmodern revolutionary' from a sophisticated and savvy use of communication technologies to carry his message worldwide.

Zapatismo represents an influential new politics of dissent that continues to resonate worldwide. In both its theory and its practice, Zapatismo is a post-ideological anarchist politics. But the Zapatistas are not anarchists, even as they draw important ideas and principles from it. Nor are they Marxists, communists, socialists or liberation theologists – as influential as these traditions have been in shaping them. They continue to resist and refuse such labels. This ideological 'refusal' is consistent with our meaning of post-ideological anarchism. The Zapatistas deny that Zapatismo is an ideology for precisely the same reasons that we have identified a post-ideological impulse within anarchism – and radical politics more generally. There is a refusal to be tied to doctrinaire, sectarian positions that are considered both authoritarian and shortsighted. A post-ideological anarchist politics is instead a fluid, inclusive and non-sectarian politics that draws on a number of positions to construct an autonomous particularistic politics from *below*. Zapatismo represents well a post-ideological anarchical politics.

6
Greening Anarchy: Social Ecology

Murray Bookchin is a radical ecology stalwart. His integration of ecology and anarchism helped create a novel form of eco-anarchism – one he prefers to call social ecology. Social ecology has contributed significantly to the greening of anarchism. Its primary insight is that the ecological crisis is a social crisis of values, with hierarchy the main culprit. Yet Bookchin claims to go beyond traditional anarchism and even newer forms of anarchism. He sees 'authentic' anarchism as highly individualist and champions municipalism, or communalism, in its stead. Municipalism describes a *polity* or civic arena in which free people participate directly in the consociational management of their community. Bookchin is certainly against the state, but not so the direct democratic practices of community self-government – a form of 'town meeting government'. To the chagrin of many fellow anarchists, social ecologists encourage the fielding of and voting for candidates in municipal elections. Bookchin also promotes a highly rationalist political philosophy which he opposes to the 'anti-rationalism' that he argues infuses much of radical ecology and now much of contemporary anarchism.

Bookchin is increasingly sceptical of the social theory and practice that passes for anarchism today. He sees in contemporary anarchist expressions – including many forms of anarchist inspired anti-capitalist activism – a capitulation to extreme individualism, postmodernist 'pastiche' and hysterical anti-rationalism. As we saw in Chapter 1, he is contemptuous of contemporary 'lifestyle' anarchism. He sees in the appeal of anarchism for many young people today, an attraction for these highly individualized, indeed narcissistic, lifestyle behaviours. Bookchin contends that anarchism's 'obsession' with autonomy has generated many 'idiosyncratic acts of defiance that verge on the eccentric' and which 'not surprisingly have attracted many young people and aes-thetes' (2003, 25). He opposes his brand of social or communitarian anarchism to the excesses of individualistic or lifestyle anarchism. He reserves his most scathing criticism for primitivists, ontological anar-chists, deep ecologists and Situationists. As a proud anarcho-socialist whose leftist roots go back a long way, Bookchin seeks to transform the social institutions and practices that reinforce hierarchy (Dryzek 2005, 207). His antipathy to deep ecology, and the vigorous debates that devel-oped between the two schools in the 1980s, rested on his charge that deep ecology was fixated on 'navel-gazing' consciousness change.

Writing since the 1950s, it was in the context of the developing environment movement that Bookchin's ideas took off. A key radical ecology school, social ecology is an innovative and multifaceted social theory that explores the impacts of domination and hierarchy for society and the environment. It chronicles the complex historical narrative of hierarchy through its various political and cultural mani-festations. It moves beyond the descriptive to the prescriptive to argue for a form of social organization that dissolves hierarchy. Since he finds no hierarchies in nature, rather differentiation and participation, Book-chin urges the application of ecological principles onto society – hence social ecology. Bookchin defines social ecology this way:

> Social ecology is, first and foremost, a *sensibility* that includes not only a critique of hierarchy and domination but a reconstructive outlook that advances a participatory concept of 'otherness' and a new appreciation of differentiation as a social and ecological desideratum. Formalised into certain basic principles, it is also guided by an ethics that emphasises variety without structuring dif-ferences into a hierarchical order. If we were to single out the pre-cepts for such an ethics, I would be obliged to use two words that give it meaning: participation and differentiation (1987, 25).

Social ecology so conceived is a specific form of eco-anarchism. As a broad umbrella term, eco-anarchism accommodates a number of ver-sions. The main ones are social ecology, bio-regionalism and primi-tivism. Bookchin's social ecology is the most prominent, although primitivism is often conceptualized as a new form of social ecology (see Morland 2004, 23). Even so, Bookchin is careful to distinguish social ecology from the broad brush of eco-anarchism, claiming that his version is far more sophisticated than 'the simplistic ideology' of eco-anarchism as a whole (2003, 27).

Bookchin also takes umbrage at the association between revolutionary syndicalism and anarchism, contending that anarcho-syndicalism offers a very limited capacity for social change. Episodic strikes and other modes of direct action are inadequate tools for revolution. For Bookchin, the Spanish syndicalists and anarchists were crippled by their incapacity to 'take the next step' in institutionalizing a workers' and peasants' democracy. This reflected their misunderstanding – one shared with the radical left as a whole – of politics as statecraft (2003, 26). Rather than fixating on 'individual eccentricities', social ecologists would instead focus on building cohesive self-directed democratic communities, a task that requires strong organizational and management skills. It is through Bookchin's communalist project that the necessary tools for building an authentic democratic politics would be honed. Communalism draws from the libertarian socialist tradition, hence its infusion in both Marxism and anarchism. It retains a socialist vision and a dialectical approach and combines it with an anti-statist and confederalist politics that respects humanity's embeddedness in the natural world. It rejects from Marxism what it sees as an inherent authoritarianism and from anarchism an extreme individualism and 'anti-rationality'.

The Institute of Social ecology was co-founded by Bookchin in 1974 and continues to promote the ideas of social ecology today. Dedicated to establishing the educational infrastructure for a free and ecological society, the Institute offers a range of educational programs, activist workshops and research activities. Both the Institute and individual social ecologists within it have been involved with and promote the anti-globalization movement (AGM). Writing for the Institute, Milstein (2004) notes that anarchism provides 'a guiding light' to today's anti-capitalist movement. The AGM takes the fight for the right to a 'free and self-governing society' to the streets of many locations throughout the world. And Bookchin's social ecology has helped inspire this dissent.

Bookchin's anarchism, however, is not necessarily a post-ideological one, even if it contains many of the 'ingredients'. He has contributed to the considerable reshaping and greening of anarchism, and promotes flexible practices such as voting, elections and 'government' traditionally abhorrent to many anarchists. But in his vigorous defence of a range of non-negotiable core principles and positions, Bookchin demonstrates a decidedly ideological bent. Even if Bookchin agreed with the conception of post-ideological anarchism, the suspicion is that he would roundly condemn it. He would probably agree that there has been a looser embrace of anarchist principles, and they have indeed been combined with an assortment of other eclectic positions. But he would arguably find little to champion in this. Instead he would glimpse in this development the death of 'real' anarchism. Nonetheless, social ecology has articulated a compelling and influential green anarchism that has 'modernized' anarchism considerably. This chapter explores the main ideas of social ecology, beginning with its theoretical and anthropological roots. It then examines social ecology's eco-philosophy, dialectical naturalism, before discussing the key political and organizational principles of the communalist model Bookchin promotes under its name.

Social ecology as eco-anarchism

Bookchin' social ecology locates the exploitation of nature in the same logic of domination that powers a hierarchical society. Like much other anarchism, social ecology's anarchism incorporates Marxist analysis but rejects its classism, statism, and authoritarianism. Unlike Marxism, social ecology is environmentally focused. In its account of social and ecological ruin social ecology draws from a number of traditions and like other radical ecology schools finds that the origins of ecological degradation are social. Its explanations for this distinguish it from the other schools already explored in the Chapter 4. But it shares with them the view that there is a very strong link between an environmental crisis and a social crisis in values. Bookchin singles out hierarchy as the root of this crisis. For Bookchin hierarchy underpins all social institutions and creates a stratified society. Furthermore, it ruptures the evolutionary propensity towards freedom and subjectivity. As an organizational principle hierarchy extends well beyond its structural boundaries and is also an internalized psychological condition, or a state of consciousness that corrodes most social and ecological relationships. For Bookchin, hierarchy 'is a sensibility toward phenomena at every level of personal and social experience' (1991, 4). This psychological condition induces human beings to practice hierarchy and domination in most aspects of social life, including their relationship with nature. This 'sensibility' towards hierarchy and domination negates the 'potential realm of freedom' in society and in nature. The claim is that the evolutionary drive towards participation and differentiation is severed by the control exerted by *some* human beings over others, and over nature as a whole.

Environmental problems thus have their roots in a social 'mutation' that has induced hierarchical practices between members of the human community and, consequently, over the natural world. Social ecology is convinced that 'the very concept of dominating nature stems from the domination of human by human, indeed, of women by men, of the young by their elders, of one ethnic group by another, of society by the state, of the individual by bureaucracy, as well as of one economic class by another' (1980, 76). This explains the crux of the term *social* to the label social ecology. According to Bookchin, the quest for freedom must begin 'not only in the factory but also in the family, not only in the economy but also in the psyche, not only in the material conditions of life but also in the spiritual ones' (1980, 76). In keeping with the anarchist legacy of unity between means and ends, social and environmental ruin can only be arrested by non-hierarchical practices. The containment of degradation through the hierarchical and centralized practices of the institutionalized state, will never yield the desired result. Rather, it will simply supplant one form of domination with another.

Social ecology draws from the tradition of critical theory. Like some other radical ecology schools, and the new social movements more generally, Bookchin took analytical cues from the Frankfurt School of critical theorists, even as he maintained his distance. In particular he singles out 'their word-magic, their defence of reason against mysticism, and their demanding intellectual level' as inspiring intellectual elements that helped guide his own analysis (1991, viii). But Bookchin's key influences are the social or communitarian anarchists, and most particularly Kropotkin. His other influences range widely, from Aristotelian entelechy, Hegelian dialectics, Marxist historicism and cultural anthropology. Bookchin is also unashamedly of the utopian tradition. His objective is to 'present a philosophy, a conception of natural and social development, an in-depth analysis of our social and environmental problems, and a radical utopian alternative ... to the present social and environmental crisis' (1991, xix). In short, Bookchin's ecological society is decidedly utopian, but it is only by ridding society of the psychology and practice of hierarchy that this utopian vision can be realized.

Disentangling Bookchin's anarchism nonetheless remains a thorny prospect. While he claims his anarchism as social anarchism and as a form of libertarian socialism, he increasingly dissociates himself from the label. Originally identifying his communalism as 'the democratic dimension of anarchism', he now claims it as a distinct ideology in itself (2003, 27). Primarily, he wishes to proclaim the democratic heart of *his* anarchism. But he is also determined to distinguish his anarchism from the 'rubbish' that passes for anarchism today – namely the offensive 'lifestyle anarchism' that he rails bitterly against. As a highly individualist anarchism narcissistically 'of the ego', for Bookchin lifestyle anarchism betrays the only authentic expression that anarchism can take: a social anarchism that recognizes that individual freedom rests fundamentally on the freedom of the community. He cites Bakunin and Kropotkin's views that the individual could never exist outside society, and that the individual's development is 'coextensive' with social development. Indeed, to speak of 'the Individual' as 'apart from its social roots and social involvements is as meaningless as to speak of a society that contains no people or institutions' (1994). Against the 'aberration' of lifestyle or hyper-individualist anarchism he poses his own highly organized and disciplined social ideology: social ecology's communalism. As the new 'green proletariat,' environmentalists – but only those of his social ecology ilk – have a distinct opportunity to lead this charge.

Despite the opportunity presented radical greens, Bookchin instead despairs of the route radical ecology has taken, especially deep ecology, and is now equally despairing of the vacuousness of lifestyle anarchism. With regard to the former, he charges that deep ecology 'has no real sense that our ecological problems have their roots in society and in social problems' (1988, 13). He laments deep ecology's implicit misanthropy and anti-humanism, arguing that to confuse humanism with a brutal anthropocentrism is to ignore humanity's wonder and uniqueness. He believes it no accident that deep ecology generates the kinds of misanthropic and 'speciest' overtones attributed to the likes of Dave Foreman from Earth First!, and Edward Abbey before him. He is equally disparaging of cultural eco-feminism's mysticism and spiritualism. These discourses' resurrection of 'nature-worship' and notions of the 'noble savage' gloss over the fact that 'ancient Egypt, with its animal deities and all-presiding goddesses, managed to become one of the most hierarchical and oppressive societies in the ancient world' (1988, 17). Such unmitigated 'nature-worship' has reduced radical ecology into a 'wilderness cult, a network of wiccan covens, fervent acolytes of Earth-Goddess religions, and assorted psychotherapeutic encounter groups' (1991, il). Even worse, its 'anti-humanism' marginalizes the movement's most powerful weapon: humanity's rich capacity 'to change the world for the better and enrich it for virtually all life forms' (1991, lx). As we have already noted, the advent of lifestyle anarchism – a development that he charges resonates the hyper-individualism and apoliticism of deep ecology and related discourses – further weakened the social change agency of radical greens.

Bookchin nonetheless derives hope from greens as a 'new proletariat' – but a hope driven primarily by his social ecology greens. Since contemporary capitalism is adept at containing crisis, and since the revolutionary proletariat has waned in numbers, class consciousness and political will, it seems that transformational hope now lies in ecological crisis (1990b, 3). His views, albeit conceptualized differently, parallel those of eco-Marxist James O'Connor's (1987; 1994) identification of environmental risk as the second contradiction of capitalism. Bookchin too argues that capitalism is producing ecological 'external conditions' of crisis. To this degree, the green movement has the capacity to become the new proletariat, the new driver of radical change. Ecological crisis provides radicals with a rallying call to arms: 'the Green movement, or at least some kind of radical ecology movement, could thereby acquire a unique, cohering, and political significance that compares in every way with the traditional workers' movement' (1990b, 3). But his 'green proletariat' is a very different proposition to Marx's proletariat. He condemns the latter as an 'undifferentiatated monolith which ... fails to act as the instrumental agent in history' (Shantz 2004, 692).

Importantly, greens are also more likely to situate their radical transformation in the community, or *polis*, and enact it through the practice of an authentic participatory politics. Unlike 'the *locus* of proletarian radicalism' which was the factory, the *locus* of the radical ecology movement would become 'the community: the neighbourhood, the town, and the municipality' (Bookchin 1990b, 3). A communitarian, ecological society would be realized through the *practice* of community, that is, through direct and active participation in the decisions that shape that community. But before we engage with the organizing principles of Bookchin's ecological society, we need to more fully examine the dominant practice his organizational form seeks to eradicate: hierarchy.

The development of hierarchy

In his major work, *The Ecology of Freedom*, Bookchin traces the development of hierarchy from pre-literate societies to its sophisticated form in the nation-state. By hierarchy he means, 'the cultural, traditional and psychological systems of obedience and command, not merely the economic and political systems to which the terms class and State most appropriately refer' (1991, 4). Bookchin stresses that hierarchy is more than a just organizational principle for stratifying and ordering society. It is also a 'state of consciousness' that is internalized and pervades all areas of social, psychological and individual experience. He traces its rise over long historical stretches that saw the anarchic or semi-anarchic organization of organic communities gradually replaced by hierarchical structures that culminated in the modern state. The egalitarian division of labour endemic to these organic communities was then displaced by hierarchies based on gerontocracy, shamanism and patriarchy.

Changed attitudes to the natural world accompanied these hierarchical developments. Organic communities that saw themselves as part of the larger community of nature gradually yielded to communities that dominated and exploited nature. The human domination of nature stemmed from these hierarchical sensibilities. These included man's domination over man, men's domination over women, the elders' domination over the young, and the domination of 'the big man' or the shaman over the community as a whole. But hierarchy was more than an externally imposed domination. Its essence was also internalized psychologically so that the idea and practice of hierarchy became 'normalized'. This was achieved by dividing 'the individual ... against itself by establishing the supremacy of mind over body, of hierarchical domination over sensuous experience' (1980, 63). Eventually, this 'objectification' of the human subject gave rise to the objectification of nature.

For Bookchin, hierarchy emerges well before the development of classes and the state, with the nation-state merely the most powerful crystallization of it. Contrary to the eco-feminist claim that the logic of domination stems from patriarchy, Bookchin contends instead that gerontocracy was one of the earliest hierarchies, even if gerontocracies paved the way for patriarchy. In order to stave off 'irrelevance and expulsion' in a world where military and physical prowess was everything, the elders compensated for their waning influence by appropriating the only sphere they could still control: the social sphere. The elders' need for social power is thus a 'function of their loss of biological power', impelling them to construct themselves as 'the architects *par excellence* of ... social power, and of its institutionalization along hierarchical lines' (1991, 81). Gerontocracy, however, is a circumscribed hierarchical power since everyone, should they live long enough, will be eventually admitted to its privilege. But in the rise of Shamanism – 'the specialists in fear' – a broader and more potent hierarchy began to emerge.

The shaman introduced more widespread practices of domination, although his power is still not systematic since the principles of 'usufruct, complementarity and the irreducible minimum' still dominated social relationships (1991, 85). Nonetheless, through his adoption of organizational and coercive mechanisms the shaman is 'the incipient state personified' and one of the first 'professionals in political manipulation'. To protect themselves should their 'magic' fail, they also begin to form alliances with others, particularly the young or the elders. The shamanic tradition thus generates the first political institutions and coalitions, centred on the maintenance and centralization of power (1991, 84–5). In addition, incipient patricentricities and patriarchies begins to emerge through the shaman's heightening of the 'masculine temperament' of a paternal community. But only when the 'blood oath' – the mutual obligations implied by blood ties – was severed could private property, independent wealth, a class society and the state emerge in earnest.

On Bookchin's reading of history, the shift from hierarchical to class societies occurred on two levels: the material and the subjective. On the material plane there occurred the development of the city and the state, the social integration of an authoritarian technics, and an increasingly sophisticated market economy. On the subjective plane – the 'epistemologies of rule' – there occurred the internalization of a command and obedience psychology. This internalized 'epistemology of rule' approximates a prefigured 'disciplinary practice', despite Bookchin's antagonism towards Foucault and poststructuralism in general. The shaman, and his incorporation into the warrior chieftains, was gradually replaced with the 'priestly corporation'. These priests were excellent promoters of individual self-discipline through guilt and renunciation. As a powerful clan in itself, the priestly corporation began to take on the shape of something quite novel, a class:

> The priesthood has the power of ideology ... that relies on persuasion and conviction. The warrior has the power of coercion. ... Hence, it was the warrior chieftain and his military companions from whom history recruited its classical nobility and its manorial lords, who produced the political state, and later, the centralised monarchy with priestly vestiges of its own. (1991, 94)

While the development of the state was a very gradual phenomenon, over time it acquired 'stability, form and identity' through institution building, the centralization and professsionalization of power and the development of law and governance (1991, 129). The state's emergence was thus specifically characterized by the gradual and historical 'polit- icization of social functions', with its culmination in the most 'supreme political act of all': the delegation of power.

At the same time as Bookchin traced a history littered with the legacy of domination, he also identifies a promising libertarian impulse running through it. It is this 'legacy of freedom' that Bookchin attempts to resurrect. Included in this libertarian tradition Bookchin locates 'the millenarian Christian sects of the Middle Ages, the Diggers' colony in the English Revolution, the town meetings in New England after the American Revolution, the Parisian sections during the French Revolution, the Paris Commune, and the anarchist communes and councils of the Spanish Revolution', as well as Athenian direct demo- cracy (Marshall 1993, 608–9). And in keeping with his historicist methodology, he concludes by stating that 'the heritage of the past enters cumulatively into the present as lurking problems which our own era has never resolved' (Bookchin 1980, 63). As a philosophy that uncovers harmony between human and human and between human- ity and nature, he posits his 'dialectical naturalism' as the way forward for resolving these lurking problems.

Dialectical naturalism

Bookchin articulates a philosophy of nature – 'dialectical naturalism' – that grounds his ethics and politics. He argues that ecological and evo- lutionary processes in nature provide the grounding for an objective ethic which in turn guides human directionality towards its potential for freedom and differentiation. His 'telos-in-nature' epistemology is teleological to the degree that it highlights a general directionality in nature as opposed to a fixed or predetermined end. Bookchin's eco- philosophy is a 'neo-organicist' one, emphasizing the 'three *attributes* of organic life that are of primary ethical significance ... mutualism, freedom, and subjectivity' (Albrecht 1993, 9). These attributes also form the basis of his conception of social and political life and inform the primary organizational principles that he advocates. For Bookchin, natural ecology is much more than the physical environment in which humanity is embedded: what 'ecology, both natural and social, can hope to teach us is the way to find the current and understand the

direction of the stream' (1991, 25). The good society becomes the ecological society, hence the neologism: eco-anarchism.

While Bookchin's social ecology draws from a variety of philosophical
and political traditions, he highlights his epistemological debt to an
Aristotelian politics and a Hegelian dialectic. Central to social ecology is
thus a teleological and dialectical philosophy of both human and non-
human nature. According to this view, natural evolution is a lengthy
spontaneous unfolding of an 'inorganic latent' realized through increasing complexity and diversity. Complexity and diversity in turn propels
increased differentiation in a seemingly unending cycle of life. In this
way, 'life itself, as distinguished from the non-living, emerges from the
inorganic latent with all the particularities it has immanently produced
from the logic of its most nascent forms of self-organization' (1991,
31–2). The central 'logic' of evolution thus becomes participation, self-
organization, differentiation and spontaneity, qualities which are also
ascribed to the social world:

> In nature, balance and harmony are achieved by ever-changing dif-
> ferentiation, by ever-expanding diversity. Ecological stability, in
> effect, is a function not of simplicity and homogeneity but of com-
> plexity and variety. The capacity of an eco-system to retain its
> integrity depends not on the uniformity of the new environment,
> but on its diversity (1991, 24–5).

It is precisely when such qualities are also assigned to the social world
in the form of an 'objective ethics' that the philosophy of social
ecology becomes particularly contentious (see Curran 1999).

The evolutionary drive towards greater differentiation and complexity, and of increasing consciousness and subjectivity, yields the
'marvellous' development of human nature ('second nature') from
its roots in non-human nature ('first nature'). Bookchin bridges the
nature/culture dualism by contending that second nature springs from,
and is rooted in, the development of first nature. Thus it becomes 'emi-
nently *natural* for humanity to create a second nature from its evolution in first nature' (1990a, 162) although Bookchin is emphatic that
this implies none of deep ecology's biocentrism. He charges that biocentrism blurs the distinction between humanity and other species by
refusing to highlight what is unique and creative in human nature.
Indeed, for Bookchin humanity is 'nature rendered self-conscious' and
this consciousness manifests in humanity's 'development of a uniquely
human culture, a wide variety of institutionalized human communities,

an effective human technics, a richly symbolic language, and a carefully managed source of nutriment' (1990a, 162). The evolution from first to second nature includes the realization of nature's potentiality for rationality and consciousness. It is this 'latent subjectivity in the inorganic and organic levels of reality that reveal[s] an inherent striving towards consciousness' (1991, 11), one that allows for 'a graded development of self-organization that yields subjectivity and, finally, self-reflexivity in its highly developed human form' (1991, 353–4). But this directedness, or seeming teleology, is for Bookchin a process of striving rather than an inevitability.

Aristotle and Hegel are Bookchin's key influences here and his dialectical approach owes them much. He nonetheless outlines a very specific reading and interpretation of their works. Bookchin adopts the Aristotelian 'entelechial' notion of causality, of the substance's striving towards the form intrinsic to its nature. His notions of dialectics and teleology also flow from the Aristotelian tradition. While Bookchin also borrows from a broad Hegelian dialecticism, especially the notion of substance as subjectivity, he rejects Hegelian idealism. He finds particularly distasteful the Hegelian notions of the 'absolute' or the 'spirit'. Bookchin instead claims that 'in an ecological dialectic ... there would be no terminality that could culminate in a "God" or an "Absolute" as Hegel does' (1990a, 169). Despite this Hegelian 'terminality', Hegel's insight on the nature of transcendence (Aufhebung) continues to influence Bookchin:

> ... the organic flow of first into second nature is a reworking of biological reality into social reality .... [where] each phase or 'moment', pressed by its own internal logic into an antithetical and ultimately a more transcendent form, emerges as a more complex unity-in-diversity that encompasses its earlier 'moments' even as it goes beyond them (1990a, 175).

Bookchin also draws his conception of 'wholeness' from the Aristotelian and Hegelian traditions even if he feels compelled to appropriately 'ecologize' it. He distinguishes 'wholeness' from the more generally accepted conception of 'holism', particularly deep ecology's version of it. He especially objects to the collapse of the term 'wholeness' into a meaningless 'oneness' (the 'holism' of the biocentric approach) since this obscures differentiation and diversity. 'Oneness' allegedly resurrects the Hegelian conception of a 'night in which all cows are black', and fails to acknowledge that 'the capacity of an ecosystem to retain its integrity depends not on the uniformity of the environment but on its diversity' (1991, 24). For Bookchin, deep ecology's 'oneness' promotes homogeneity rather than diversity. This simplification of diversity and complexity in nature is then reflected in the decreasing diversity and complexity in culture and nature. Just as the natural world is simplified and homogenized, so too is society and subjectivity. Furthermore, deep ecology's attempts to heal the rift between human and non-human nature only succeeds in accelerating a ruinous reductionism, one that encourages 'an intellectual dissolution of *all* difference into an undefinable "Oneness" ... that turns a concept like "interconnectedness" into the bonds of a mental and emotional straightjacket' (1991, xlvi).

Bookchin's caustic language has intensified the deep ecology suspicion of what it considered his latent anthropocentrism. They charge that in his championing of a unique human capacity for rationality can be found an implicit human chauvinism. Indeed, Bookchin argues that 'whether we truly know and fully appreciate first nature depends very much on having the intellectual and emotional ability *not* to confuse ourselves as human beings with coyotes, bears, or wolves, much less with insensate things like rocks, or rivers, or even more absurdly, with the "cosmos"' (1991, 1). Of course, there are many counter-arguments that deep ecologists have levelled and continue to level against these assaults. The rancour between Bookchin and other radical ecologists, not just deep ecologists, continues relatively unabated especially now with his attack on lifestyle anarchism. The accommodation between Foreman of Earth First! and Bookchin in the late 1980s was relatively shortlived, especially in the face of more recent disputes (see Bookchin & Foreman 1991).

Bookchin's main complaint was deep ecology's refusal to subscribe to a central insight of social ecology – its conception of unity-in-diversity. A central Hegelian idea, unity-in-diversity underpins social ecology. Once again, Bookchin rejects the Hegelian 'Absolute' and Hegel's implicit authoritarianism and statism, positing instead that the ecological society drives humanity towards increasing rationality, freedom and subjectivity. However, their views coincided in so far as Hegel views freedom and ethics as essentially social in character, and that the individual's identity is thus socially circumscribed. For Bookchin, along with the 'need for the human being for independence, and to be separate, and autonomous, is the need for unification with others ... to become part of the social whole' (1987, 244). This conception of a 'social whole' corresponds to the Bookchin insight that the individual's autonomy depends on its mutualistic, associative and collaborative character.

A final influence on the philosophy of social ecology is Hegelian historicism, especially as outlined in Hegel's *Phenomenology of Mind* (1969) and *Reason in History* (1953), an influence Marx also lays claim to. According to the historicist view, social development relies on an understanding of society's history, roots and practices. Bookchin's major work, *The Ecology of Freedom* (1991) adopts a historicist structure in its exploration of the roots of hierarchy and the consequent erosion of freedom. Importantly, however, in merging historicism and dialecticism, Bookchin does not simply apply a dialectical method to analysing society. Rather he claims to find a dialectical process *in* society, even if Bookchin roots this process in nature to a degree never contemplated by Hegel. We now turn to what is perhaps the most contentious area of his eco-philosophy: his claim that nature provides an objective grounding for ethics.

Nature as a grounding for ethics

For Bookchin, hierarchy disrupted the propensity for freedom and subjectivity. Its main manifestations – patriarchy, gerontocracy, patricentricity, racism, colonialism and imperialism – have set dangerous precedents for the operation of society. But for Bookchin there is also in place a 'redemptive dialectic' where hierarchical society's power to destroy can be replaced by the power to create (Marshall 1993, 610). He argues, contentiously, that it is nature itself that provides the map or guidelines for the organization of social freedom. In short, nature provides the grounding for an ethics of social life that restores it on the path of freedom. But Bookchin distinguishes between nature as the *realm* of ethics and nature as a *grounding* for ethics. He claims that nature is not in itself an ethics, but rather provides a '*matrix* for an ethics'; it is a 'source' rather than a 'paradigm' and 'a *ground* from which to creatively develop ethical ideals' (1991, 278). Nonetheless, this matrix for ethics opens up the possibilities of a 'third' nature or a 'free' nature so that human nature becomes 'more wilful, self-reflexive, and consciously creative' (1987, 74). An ecological society thus actualizes the evolutionary potential of freedom and subjectivity. At the same time, Bookchin is aware of the significant limitations of 'nature philosophies' that posit nature as a grounding for ethics; his criticisms of deep ecology are in part based on this. More alarmingly, the 'blood & soil' justification for fascist ideologies is also well documented. Nonetheless, he does not dismiss them altogether, charging that 'nature philosophies' also have the potential to yield freedom and cooperation.

His major claim is that an ecological society resurrects the thrust towards mutualism, consociation and cooperation – qualities embedded in both non-human and human ecology. Here Bookchin draws from Kropotkin who also found in nature propensities towards cooperation as well as competition. In one of his most famous works, *Mutual Aid* (1939), Kropotkin argues that the most successful species maintains their survival against external threats through a resilience built on cooperation and mutualism. For Kropotkin, mutualism is thus one of the most important factors of evolution:

> ... under *any* circumstances sociability is the greatest advantage in the struggle for life. Those species who willingly or unwillingly abandon it are doomed to decay; while those animals which know best how to combine have the greatest chances of survival and further evolution (1939, 60–1).

Unlike Darwinists, or more specifically Social Darwinists, Kropotkin argues that the struggle for survival manifests more so against external forces than within species themselves. The most durable species are those that successfully cooperate and unite against these external forces. It is those animal species 'in which individual struggle has been reduced to its narrowest limits, and the practice of mutual aid has attained the greatest development' that survival is more likely to be guaranteed (1939, 230). Kropotkin included human beings among these animal species thus applying his own ecological principles to social life. While not agreeing with all Kropotkian analysis, Kropotkin's mutualistic naturalism significantly informs Bookchin's own nature philosophy.

From these ecological observations Bookchin concludes that an 'ethics of complementarity' drives social life. This ethic arises from the interdependency that underpins the initial relationship between parent (especially mother) and child, an interdependence that then radiates onto other social relationships. For Bookchin, 'the cradle of social life' stems from the extended dependence of the human child on its parents. Human nature and human attributes are thus 'a biologically rooted process of consociation, a process in which cooperation, mutual support, and love are natural as well as cultural attributes' (1991, 317). This process fosters a sense of *inter*dependence as opposed to a narrow, egoistical *in*dependence. Human nature is shaped by the mechanisms of an organic, and biological, process that reaches into all areas of social life:

> The prolonged process of physical maturation in the human species turns individual human nature into a biologically constituted form of consociation. Indeed, the formation not only of individuality but also of personality consists of being *actively* part of a permanent social group ... Without the care, cooperation, and love fostered by the mother-child relationship and family relationships, individuality and personality ... begin to disintegrate, as the modern crisis of the ego so vividly indicates (1991, 317).

Thus, human survival is guaranteed by 'a maternally biased need to associate, to care for one's own kind, to collaborate' (1991, 343). By externalizing this ecological self-organizing principle, authoritarian structures such as the state sever the consociation and complementarity of social life. Strategies like direct action become, therefore, not only practicable strategies for achieving practicable ends, but are also a human reclamation of evolutionary rationality. In organizing their own lives and their own societies, humanity fulfils its ecological potential.

Bookchin's ecological and anthropological studies claim to vindicate an objective ecological ethics that can be directly applied to social life. But notions of objective ecological ethics are, of course, highly contentious and have been extensively criticized. As one critic asserts, the development of 'a moral imperative from an empirical observation, an "ought" from and "is"' runs 'the logical risk of the "naturalistic fallacy"' (Marshall 1993, 610). While Bookchin is cognizant that his theory risks such a charge, he is resolute in his defence that nature provides a *grounding* for ethics. He acknowledges that 'first nature remains a realm of ethical vacuity' and that what distinguishes humanity from other living organisms is its ability to create an ethics: first nature is never 'cruel' or 'kind', 'heartless' or 'caring', 'good' or 'bad'; ethics, rather, 'was *born* with human society, just as surely as metabolism was born with the emergence of life' (1991, xxxv). But whether this defence satisfies critics is another matter altogether (see Eckersley 1989).

Central to Bookchin's objective ethic is humanity's relationship to nature. Humanity relates actively rather than passively to nature. A meaningful mediation with nature arises through humanity's labour and their creation of 'technics'. Humanity does not simply live in the natural world, it also transforms it. Nature provides the means of abundance and with it a propensity for human freedom. Human political striving seeks to transcend material scarcity in the process of creating a 'post-scarcity anarchism' characterized by a mutually enjoyed abundance. Marx too highlighted the necessity of material abundance in realizing a free society. But for Bookchin, like many anarchists, it was not just capitalism but hierarchy in general that privatized abundance and corralled it into fewer and fewer hands. Like Marx, Bookchin sees much scope in technology and, compared to many radical ecologists who are more circumspect, he is a technological optimist. This helps explain his disparagement of anti-technology anarcho-primitivists. Bookchin does not blame technology for ecological crisis, arguing that ecological ruin lies more in the ownership of the technology rather than its utilization. He thus opposes 'libertarian' technics to 'authoritarian' ones – terms he borrows from Lewis Mumford (1991). A libertarian technics develops in a social and ethical context that seeks to create tools of abundance, whereas an authoritarian technics simply seeks the honing of the tools of hierarchy. He suggests that it is when social and ethical contexts are stripped from the technical, that an authoritarian and destructive technics develops. Thus it is not 'industrial productivity that creates mutilated use-values but social irrationality that creates mutilated users' (1991, 70).

Overall, there remains a good deal of problematic analysis in social ecology's eco-philosophy. Difficult to confirm or deny, Bookchin presents an 'ecological image' of nature that primarily 'works as a metaphor' (Marshall 1993, 619). Another critic notes that '[w]hen Bookchin claims that his ethics is objective, he means that it is based on potentialities that are actually latent in nature, potentialities that really exist as concrete possibilities standing beyond the present' (Eckersley 1989, 106). A non-ecological society may interfere with a dialectical process that seeks to actualize of latent potential, but there is still no guarantee that, even if actualized, this potential is intrinsically ethical and automatically yields freedom. While for Bookchin the evolutionary thrust is one towards increasing freedom and subjectivity, his notion of freedom is so broadly constituted it risks typecasting both humanity and nature:

In drawing a parallel between the developmental path of an acorn, a human embryo, nature, and finally society – as if all have an equally discernible objective standard of fulfillment [for example, 'freedom'] – Bookchin is collapsing ontogenetic development (i.e., the sequence of events involved in the development of an individual organism) into phylogenetic evolution (i.e., the sequence of events involved in the evolution of the human species, including its culture) (Eckersley 1989, 106–7).

These conceptual problems may simply reflect Bookchin's failure to provide clear definitions of some of his central terms, especially of concepts such as diversity, spontaneity, harmony, freedom and stability. His generic application of these terms to describe both non-human and human processes thus invites circumspection. His conception of diversity is a case in point. He provides little guidance for distinguishing between exemplary and 'aberrant' forms of differentiation. When he states that 'the greater the differentiation, the wider is the degree of participation in elaborating the world of life' he provides few tools for assessing the *quality* of the participation being applauded. After all, 'the same words are used by functionalist sociologists and system theorists, as well as ecologists' (Marshall 1993, 620). Perhaps in response to such criticisms of his innovative but problematic dialectical naturalism, Bookchin has more recently turned to honing his *political* response to ecological crisis: his conception of the 'communalist project' (2003).

Municipalism: the communalist project

Direct participation and a grassroots politics are the main strategies for the realization of Bookchin's ecological society. They are also the means through which his anarchist unity between means and ends are met. The direct unmediated politics of the municipality not only bypasses the state's centralized form but also restores individual and community autonomy and self-reliance. In short, Bookchin's 'brand' of anarchism, 'favors management, plans, and regulations formulated democratically by popular assemblies, not freewheeling forms of behavior that have their origin in individual eccentricities' (2003, 27). Once again, this brand constitutes its own ideology: the ideology of communalism which he argues favourably distinguishes his eco-anarchism from all other comers. Communalism is not simply a 'mere variant of anarchism'; it demands far stronger organizational structures than traditional anarchism would permit, especially in its championing of some forms of voting, elections and institutional forums – indeed of some government, albeit one exercised by citizens. Bookchin is happy to run with a dictionary definition of communalism as 'a theory or system of government in which virtually autonomous local communities are loosely bound in a federation' (in Bookchin 2003, 28).

Bookchin's communalism sits in the libertarian socialist tradition. It combines libertarian municipalism and dialectical naturalism and draws from 'the best of the older Left ideologies', Marxism and anarchism (2003, 27). From Marxism he extracts a rational and coherent socialism that 'integrates philosophy, history, economics, and politics'; from anarchism, he draws its commitment to anti-statism and confederalism as well as its overarching critique of hierarchy (2003, 27). Despite the limitations of these traditions, Bookchin argues that they contain core requirements of an ecological society: direct participation, decentralized communities, an autonomous politics and a free and active citizenry. It is one thing to oppose statecraft, which like all anarchists he does, but quite another to propose a politics to replace it. For Bookchin, contemporary anarchism, particularly an individualist, lifestyle anarchism, is characterized by its *lack* of politics, since politics '*is* the active engagement of free citizens in the handling of their municipal affairs and in their defence of its freedom' (2003, 26). True politics can only take place in a *polis* or community that is organized around the practice of direct and unmediated decision making. It is a polis unsullied by anarchist individualism or anti-rationalism, nor Marxist authoritarianism and vanguardism. It relies not on the industrial proletariat as the main historical agent but rather on a popular democratic assembly that is both green and intensely political.

For Bookchin, direct action and direct participation become both the means and end of an ecological municipalist society. But direct action is more than a strategy, it is also the means of 'ethical character building in the most important social role that the individual can undertake: active citizenship' (1991, 339). As we saw, the sociability that springs from natural evolution underwrites humanity's essentially social character. In Kropotkian terms evolution provides humanity its strong drive towards mutualism, self-organization and freedom, a drive disrupted by hierarchy. A social life not circumscribed by hierarchy yields the qualities of association, collaboration and cooperation. These are the very qualities contained in the free, unmediated democracy that Bookchin claims his municipalist community to be. It is also the point at which he champions the virtues of the (albeit limited) electoral process:

> In marked contrast to the various kinds of communitarian enterprises favored by many self-designated anarchists, such as ... food coops, and backyard gardens ... adherents of Communalism mobilize themselves to electorally engage in ... the municipal council ... and neighborhood assemblies (2003, 31).

But, once these assemblies assume some level of decision-making power they would use it to 'delegitimate' the statist institutions and organizations that currently control society. In a form resonating Proudhonian federalism, these autonomous assemblies would then unite into a highly charged confederalist model:

> Once a number of municipalities are democratized along communalist lines, they would methodically confederate into municipal leagues and challenge the role of the nation-state and, through popular assemblies and confederal councils, try to acquire control over economic and political life (Bookchin 2003, 31).

The influences of Athenian democracy are clearly evident. Bookchin suggests that 'the entire Athenian system was organized to obstruct political professionalism, to prevent the emergence of bureaucracy, and to perpetuate an active citizenry *as a matter of design*' (1991, 131). Thus was born the ideal of direct democracy – a model that animates Bookchin's organizational design. Despite its embeddedness in a slave culture and a patriarchal structure, this fledgling Athenian democracy also carried with it the seeds of freedom and democracy. The organizational principles of the Athenian assembly are utilized by Bookchin in structuring his own confederalist model:

> It is above all a network of administrative councils whose members or delegates are elected from popular face-to-face democratic assemblies, in the various villages, towns, and even neighbourhoods of large cities. The members of these confederated councils are strictly mandated, recallable, and responsible to the assemblies that choose them for the purpose of coordinating and administering the policies formulated by the assemblies themselves (1992, 297).

Confederalism would thus restore the political into the centre of social life where Bookchin believes it rightly belongs.

In keeping with the Athenian conception of *polis* and political citizenship, confederalism would reverse the link of politics with statecraft. Instead of a passive constituency whose only 'political' role is that of elector, it would instead promote an active and directly participative community. An important Hellenic contribution was the fusing of ethics and politics in the quest for 'the good life'. Bookchin concurs that politics properly understood is a 'participatory dimension of societal life and the activity of an entire community' (1992, 227). Thus, the organizational principles of an ecological society are ones that effectively distinguish politics from statecraft, policy making from administration, 'rugged individualism' from 'authentic' individuality, dependence from *in*dependence, and finally independence from social *inter*dependence (1992, 253).

Overall, an ecological municipalist society requires the activation of a number of 'coordinates' or structural mechanisms in order for it to function (1992, 257–64). First, the 'citizen's assembly' needs to be revived. This would embrace an organizational form that restores face-to-face decision making on a scale in which all citizens are able to directly participate. Second, these assemblies would need to instigate a 'confederate relationship' where they could effectively coordinate and communicate with each other. New communication technologies can assist this process. Third, municipal democracy must incorporate an educative process that promotes 'the values of humanism, cooperation, community, and public service in the everyday practice of civic life' (1992, 262). This 'etiquette of civic behaviour' would oversee the free exchange of, and respect for, difference and diversity. To this extent it would be a communicative democracy as well as a creative environment in which civic affairs and civic solidarity are given prominence. Finally, a true economic democracy must be created, one in which 'municipalization brings the economy as a whole into the orbit of the public sphere, where economic policy can be formulated by the *entire* community' (1992, 263). Bookchin claims that many of these principles are also contained in the affinity group model that he promoted long ago. The fact that the affinity model is now one enthusiastically embraced by oppositional politics and the anti-globalization movement as a whole, testifies to the embrace of his vision.

The achievement of a confederated municipalism nonetheless requires 'contravention' of some key anarchist 'dogmas'. Once again, Bookchin unashamedly pronounces his departure from anarchism: communalism 'decidedly calls for decision-making by majority voting as the only equitable way for a large number of people to make decisions' (2003, 31). Furthermore, municipalists happily run candidates in local elections. The hope is that once elected these municipalists would begin agitating and legislating for the more widespread use of popular assemblies. In turn these popular assemblies could ensure the creation of effective forms of 'town-meeting government' (2003, 33). Municipal governments that are run 'authentically' and democratically by 'free' citizens would prove combative opponents to a statist culture intent on both maintaining and expanding its power. Successful 'revolution' requires the opposing of statism's organizational as well as political form.

Conclusion

Social ecology is clearly a complex, comprehensive and systematic account of the roots of ecological crisis. While Bookchin proclaims his reluctance to outline a blueprint for a future society, acknowledging the authoritarian dangers of political blueprints, he is far from reticent in doing so. He outlines a sophisticated historical account of the rise of hierarchy and domination to which he attaches an equally reflective account of the link between nature and culture. Convinced that ecological crisis is rooted in a social crisis of values, and armed with his libertarian socialism and dialectical naturalism, he outlines the organizational mandates of his democratic municipalism. It is the democratic dimension of anarchism that impels Bookchin most. His municipalism represents the application of direct democracy even in a complex, populous society. He overlooks the fact that his municipalist prescriptions might work better in small, pastoral and decentralized environments such as those of his home state of Vermont in the United States. But most of his political and organizational prescriptions are set on overthrowing the hierarchical sensibilities that drive social and environmental ruin. And his fundamental argument that until there is a harmonization of human with human there cannot be a harmonization between humanity and nature, remains a compelling one.

What many critics have found less compelling is not so much his message as its 'delivery'. As we noted throughout the chapter, it is the combative and highly charged *delivery* of his position which has alienated many. While a source of considerable inspiration, Bookchin's social ecology raises considerable consternation even among anarchist sympathizers. His blanket condemnation of any other 'versions' of anarchism, ecologism or radicalism, and his unrepentant technological optimism, risks charges of the very dogmatism he charges others with. This is exacerbated by the language he utilizes to cast his criticisms; for example:

> Alas, we are witnessing the appalling desiccation of a great tradition, such that neo-Situationists, nihilists, primitivists, antirationalists, anticivilizationists and avowed 'chaotics' are closeting themselves in their egos, reducing anything resembling public political activity to juvenile antics (1994).

Not unexpectedly, he invites equally vitriolic counter-charges, or perhaps even worse, an increasing marginalization. This marginalization is more often a response to his perceived authoritarianism than to the force of his analysis, which continues to inspire many. To this degree, Bookchin's sectarian contrariness inclines his anarchism towards an ideological rather than post-ideological one, even though his social ecology contains many post-ideological anarchist elements.

7
Reclaim the Streets

Reclaim the Streets (RTS) enacts the holy grail of anarchism: unity between means and ends. In reclaiming space from an avalanche of capitalist encroachments, it creates a 'temporary autonomous zone' (TAZ) and a politics of 'pleasure' that celebrates identity, creativity and autonomy. Its radical ecology roots influenced its original conception as an anti-roads and anti-car movement, but it was to become much more than this. One of its prime challenges was to the dominance of cars in urban streets. But the car culture it opposed was emblematic of how capitalism colonized public space by corralling its use and curtailing its function. RTS wanted to return the public spaces consumed by the car culture back to the communities it rightly belonged to. Employing Situationist ideas and strategies, it sought to 'subvert the dominant paradigm' by counterposing starkly oppositional activities – dancing and partying – to those of the sombre car and business culture. In this way it resisted and challenged a globalization that imposed a monocultural blueprint of the 'good life'. In liberating public space from not only cars but also from the encroachment of a hollow materialism into all reaches of life, RTS represents a modern-day anti-enclosure movement. The fact that RTS events are full of colour, music, dancing and merriment should not overlook their identity as serious pieces of political theatre. Through flagrantly oppositional activities, they challenge the corporate paradigm and reveal that 'another world is possible'.

RTS is both an organization and a tactic. Its direct action strategies are a deliberate rejection of mainstream mediated politics. Demonstrating its anarchical credentials, it challenges the increasing surrender of autonomy to an external state that shrivels the capacity for independence and self-direction. It claims that immersion in direct action resurrects autonomy and hence the individual's integrity. RTS's temporary autonomous zones consciously establish areas free of, and outside, state control. Direct action resistance is much more than a tactic of 'street liberation'. It represents both the reclamation and doing of autonomy for the individual and the community against a state-circumscribed being in the world, or on the street. A RTS street party thus becomes a transitory theatrical enactment of a social alternative that subverts hierarchical power by temporarily dissolving it. It is a process whereby the street party – *to* street party – is transformed into a verb. According to London RTS (1997, 6):

> To 'street party' is to begin reconstructing the geography of everyday life; to re-appropriate the public sphere; to rediscover the streets and attempt to liberate them. To 'street party' is to rescue communality from the dissection table of capitalism; to oppose the free market with a vision of a free society.

There is strong resonance between the philosophy of RTS and anarchist impulses. RTS illustrates and exemplifies well a post-ideological anarchist politics. This resonance goes beyond a shared anti-capitalism and a shared concern with social and ecological ruin. It is in RTS's recognition that the objectives of a just, free and ecological society can only be achieved through means matching these ends, that the resonance is at its strongest. The temporary carving out of a space free from external control – a temporary autonomous zone – thus underpins RTS politics. Both RTS and anarchism place the dissolution of hierarchical power relations at the centre of a direct action template that has the reclamation of autonomy as its basis. In considering these themes, the chapter begins by exploring the origins of RTS, tracing it from its roots in Earth First! and the anti-roads movements to its tactical centrality in the anti-globalization movement (AGM) as a whole. In keeping with a post-ideological anarchism constituted of a range of elements, we also examine some of the other influences, particularly Situationism, that shape the politics of RTS. As RTS's key political strategy, the chapter examines the street party model of dissent, before closing with a fuller consideration of RTS's anarchist temperament.

The origins of RTS

RTS is closely connected to Earth First!, the subject of Chapter 8. Earth First!'s commitment to the strategy of direct action and its underlying distaste for the car culture inspired the formation of the anti-roads movement. As both strategy and principle, the anti-roads movement was influential in the formation of different chapters of RTS. While the origins of RTS is contested, there is general agreement that it emerged in London in the early 1990s out of a combination of influences from Earth First!, the rave scene and loose collections of radicals (Doherty 1999; Wall 1999; Boyd 2002). Others contend that RTS was formed by Earth First!ers in 1992 to combat the devastation of nature in the cities – hence its anti-car culture focus (Do or Die Editorial Collective 2003, 17). Either way, the connection between RTS and Earth First! is a very close one. According to London RTS (1997) RTS was originally formed in London in 1991 around the same time as the first chapter of the British Earth First! and the anti-roads movement. The anti-roads movement was gearing up to challenge the Thatcher Government's 1989 transport policy, *Roads for Prosperity,* which prioritized the building of new roads despite widespread environmental concerns. One of the earliest anti-road protests was against the extension of the M3 at Twyford Down in 1992. Inspiring further protests, Twyford Down provided an emerging template for future anti-road actions. These protests were generally organized by two key players: a local community group and a group of 'outside' radical ecology protestors practising the tactics of direct action (see Doherty 1999, 276). It was through these anti-roads protests that radical ecologists began to penetrate both the anti-roads and anti-globalization movement in the United Kingdom (Doherty 1999).

It was also during this period that RTS consolidated its identity, becoming both an event and a strategy. RTS events, or parties, were soon a global phenomenon, held in many cities and towns throughout the world, albeit primarily in the north. RTS parties were either held on their own to protest against specific community concerns, or in collaboration with larger protests such as anti-globalization events. RTS actions thus vary in size, frequency and expression. As the RTS tactic captured the protest imagination, its form and ethic began to change. Its British origins as first environmental protest and then a site of working class struggle – when RTS came out in vigorous support of striking dockworkers in Liverpool – was gradually transformed into a more active anti-capitalist and anti-consumerist movement. It began assuming the activist ethic of organizations such as Adbusters who in turn had been influenced by groups such as the Yippies and particularly the Situationist International (Hirsh 1999; Klein 2001, 345–60).

The action at Tywford Down was a defining 'moment' of resistance. It not only captured considerable attention, but also announced RTS's coming of age as a contemporary spectacle of dissent. While Tywford Down was primarily an environmental action against the ecological impacts of road building, London RTS noted that the M11 campaign of 1994 highlighted wider social and political issues: '[b]eyond the anti-road and ecological arguments, a whole urban community faced the destruction of its social environment with the loss of homes, degradation to its quality of life and community fragmentation' (1997, 1). The ensuing Criminal Justice and Public Order Act of late 1994 imposed stronger criminal sanctions on civil protest and specifically outlawed raves, or unlicensed public parties with sound systems. Rather than containing it, this had the effect of radicalizing the movement further. When the M11 and Claremont Road campaign was lost, RTS was reformed and revitalized. The series of super-charged street parties that resulted captured significant media attention and cemented this direct action strategy as exemplary resistance politics. The street party as theatrical politics quickly crossed the British border and spread to many other cities across Europe and the globe.

While still anti-car in principle, the focus of RTS moved more rapidly to the car culture's political underpinnings. For RTS the main objective became to dissolve the power of capitalism itself:

[C]ars are just one piece of the jigsaw and RTS is about raising the wider questions behind the transport issue – about the political and economic forces which drive 'car culture'. Governments claim that 'roads are good for the economy' ... [I]t is all about increasing 'consumption', because that is an indicator of 'economic growth'. The greedy, short-term exploitation of dwindling resources regardless of the immediate or long-term costs. Therefore RTS's attacks on cars cannot be detached from a wider attack on capitalism itself (London RTS 1997, 2).

But this wider attack would not be burdensome and tedious. It would instead be pleasurable. RTS refused to measure their impact in traditional terms of protest wins or losses; they would evaluate success in a much broader way. While a prime objective of RTS and anti-roads was to prevent the destruction of the natural environment, the experience of dissent would be counted as equally important.

To be successful an action had to also 'reaffirm life' and activate pleasure through the reclamation of autonomy. This autonomy and individuality would be significantly different from materially-circumscribed ones, however. Referring to the ultimately unsuccessful M11 Claremont

Road occupation in 1994, Wall (1999, 78) notes that it was 'a riot of colour, packed with murals, sculptures made from distorted waste, a giant chessboard, a spider's web of nets stretched across from roofs and a 100-foot high scaffold tower'. In this environment art became a weapon and creativity was difficult to distinguish from strategy, since creativity *was* strategy. The description of a similar event in Toronto in 1998, captures well the meaning of 'organized chaos':

> The streets were covered in chalk, and the intersection ... had been shut down for a good hour or two. Drums and dancing were heating up an already hot mid-May afternoon. There were people from community groups performing street sculptured plays, and urban-guerrilla types planting new flowers ritualistically in the middle of the road ... there was not a car in sight ... A small part of the city had been reclaimed by its residents ... What seemed to matter was the celebration of community in the face of concrete conformity that shapes normal urban living (Hirsh 1999, 1).

In short, an autonomous space of resistance, creativity and community had been carved out from an otherwise hostile corporate environment. With it, the street party as autonomy writ large was launched in earnest. These parties were 'both a negative act of resistance and a positive act of celebration, community building and self expression' (Boyd 2002, 247). Because they were relatively easy to organize, these festivals of resistance soon expanded. By the time of the World Trade Organization meeting in Seattle in 1999, RTS-style protest had emphatically 'taken hold of the activist imagination' (Boyd 2002, 247).

The street party

Central to the street party is its identity as a Do it Yourself (DIY) event. While there are organizers who arrange the ingredients of the party – the date, the venue, the music, the activities and so on – the character of the event is determined by the participants themselves. Participation sits at the centre of the DIY ethic. Since a DIY event is only as successful as its participatory input, it is both defined and animated by participation:

> Reclaim the Streets is a protest that only works if everyone participates. This is true not only for the organisers who have to create sound systems, train with tripods, build props and compose info

sheets, but also for those who just show up on the day of the protest in costume, with radios, drums, or fire-breathing apparatus, and ready to dance ... what happens at the action depends upon what people bring with them and what they do once they are there (Duncombe 2002, 220).

With participation its underpinning ethic, the RTS event is spontaneous, autonomous and diverse, characteristics central to its anarchical temperament. The street party's organizational form welcomes all participants freely since there is no requirement for specialist knowledge or training. Critical Mass participants – groups of cyclists peddling to the event from local or distant locations – often join RTS events, especially during the larger anti-globalization protests. They not only add considerable mass, but also showcase transport alternatives in the event's symbolic mix. According to Klein (2001, 351), RTS is in many ways 'the urban centrepiece of England's thriving do-it-yourself subculture', a form now exported to the many other countries.

Organization and tactical coherence are nonetheless important. To counter expected police resistance, the venue of the street party is kept secret until the very end, often the very day of the RTS event. This reflects the organization of the original 'raves' from which RTS draws much of its inspiration. Event participants usually gather at a pre-ordained place and then depart together to the secretly designated street party site, a site revealed only to a small organizational coterie. The music system's clandestine arrival at its secret destination is usually timed to coincide with some theatrical 'stunt' designed to block traffic and capture public attention. Some of the dramatic stunts in Australian RTS illustrate well the theatricality of the event. To blaring techno-sounds and blazes of colour, participants in a Sydney RTS:

> Blockaded the streets to traffic with three huge bamboo tripods, erected a bizarre art installation sound tower pumping out psychedelic dance music, built a permaculture garden in the middle of the road and had an all day street party in the liberated zone – dancing, playing street cricket, reading the Weekend Papers ... (Luckman 2001a).

Describing 'Street Party 3', Wall (1999, 87–8) also notes its intense theatricality:

> [It was] remarkable in terms both of the nature of the territory and of the number of participants. In July 1996, 7,000 people occupied a

stretch of motorway in west London in an action initiated by RTS ... The event was planned with great care, and less than ten individuals knew the ultimate location ... participants were redirected from a meeting point at Liverpool Street station, east London, via the District (underground) Line. The policy presence was too small to prevent the street party, which mobilised more individuals in illicit activity than has any other anti-road direct action protest event.

The street party draws from historical moments of resistance and rebellion and is determined to maintain this tradition. RTS notes that many revolutionary actions – the storming of the Bastille, the Paris Commune and the 1968 uprising – were organized as boisterous popular festivals or carnivals. But it was the principles that underpinned these festivals of resistance that most animated RTS:

> Crowds of people on the street seized by a sudden awareness of their power and unification through a celebration of their own ideas and creations. It follows then that carnivals and revolutions are not spectacles seen by other people, but the very opposite in that they involve the active participation of the crowd itself. Their very ideas embrace all people, and the Street Party as an even has successfully harnessed this emotion (London RTS 1997, 3).

This then became the *raison d'être* for the street party: the enacting of a spontaneous, autonomous and participatory politics that mirror the objective of a free society. The street party sets out to counter-appropriate the practice of pleasure from a market culture that has privatized it. Against the marketization and commodification of the 'means of entertainment', the street party liberates the capacity for pleasure by reclaiming it. A free street party that creates DIY entertainment challenges the privatization or 'enclosure' of fun. But creating a liberated space within a street represents more than a simple act of autonomy. According to London RTS (1997) a commodified culture seeks to 'keep everyone indoors', separated and atomized, and confined to buying individual rather than shared pleasures. The street however provides liberation from social confinement and materialist myopia:

> The street, at best, is a living space of human movement and social intercourse, of freedom and spontaneity. The car system steals the street from under us and sells it back for the price of petrol. It privileges time over space, corrupting and reducing both to an obsession

with speed or, in economic lingo, 'turnover' ... The privatisation of public space in the form of the car continues the erosion of neighbourhood and community that defines the metropolis ... Community becomes commodity (London RTS 1997).

RTS parties and actions are non-violent, seeking to communicate their points and principles in direct but peaceful forms, even if it does not always work out this way. Many street events have been marred by vandalism and confrontations with police (Monbiot 2000). The parties' lack, indeed antipathy, to central control can mean that some individuals or groups gain a temporary control over the events' organized chaos and turn them confrontational. Resistance to centralized, co-ordinated 'crowd control', leaves the party vulnerable to 'outsiders' who do not necessarily share the same protest values. The openness that underpins RTS events can thus be a two-edged sword. RTS attracts a broad array of participants, albeit mostly young people. Among them are groups, including some ideological anarchists, who exploit the opportunity to participate in more combative ways. Intent on provoking violent confrontations in retaliation for the violence of capitalist society itself, these anarchists take full 'advantage of the opportunity to vent their fury on banks, jewellery shops and local branches of McDonald's' (Klein 2001, 349). Not unexpectedly, such developments attract more media attention than the less dramatic, and less newsworthy, peaceful events.

Not all RTS events are street parties, however, and can assume a number of forms. Some of the most successful RTS events have involved 'guerrilla gardening' – an endeavour that goes back a considerable way. While not a street party as such, guerrilla gardening borrows extensively from the toolkit of street politics to enact occupations of streetscapes and celebrations of alternative public spaces. Guerrilla gardening showcases the action diversity of street politics. Planting vegetables, trees, flowers and grass, these community gardens represent 'an attempt to address the lack of green spaces' in the city and 'to engage with the local community' in more meaningful ways; since the garden is 'a social space', it provides the opportunity 'to engage with other like-minded people' (in Millar 2005, 11). Community gardening invites people to emerge from their isolated and fortressed living spaces to interact with others in a collaborative and creative way.

Green guerrillas had operated successfully in New York for several decades where they helped transform vacant city lots into vibrant community gardens (Millar 2005). During New York's urban decline in the 1970s, there was many 'reclamations' of derelict buildings and vacant blocks, alongside guerrilla gardening activities. These activities often attracted the blessing of local authorities in recognition of their beautification effects on the city landscape. For many years city authorities and activists lived relatively harmoniously (Duncombe 2002). The East Village, already an alternative and bohemian community, became a hive of garden activism. All was to change with an economic 'revival' in the 1990s that spelt the decline of community. Determined to transform the city, incoming New York mayor Rudolph Guiliani enacted stringent law and order measures to 'clean up' the city. He accompanied these measures with an economic revival that accelerated the privatization and gentrification of the city's urban spaces. But it was not only the green gardeners and community activists who experienced marginalization in this new economic environment. Small business, once considered the heart of commerce in New York, was increasingly pressured by the influx of the larger corporations it was unable to compete with. As a result, the 'Do-It-Yourself spirit' of small business in New York was quickly 'replaced with the corporate service mantra of We-Do-It-For-You' (Duncombe 2002, 220). Interpreting this neo-liberal assault as a direct attack on their freedom and their livelihoods, these diverse New Yorkers began organizing in earnest.

As increasing numbers of previously 'communal' urban blocks were sold off in a frenetic real estate market, and economic competition proceeded apace, outraged community residents, including environmentalists and political activists, formed a broad coalition set on fighting these developments (Duncombe 2002, 223). Ideologically compatible, these community campaigners soon joined forces with the growing band of RTS activists to forge a formidable alliance of dissent against the burgeoning neo-liberal encroachment of community space. Proclaiming that if there 'is no place to freely assemble, there is no free assembly. If there is no place to freely express, there is no free expression', New York's first RTS organized around the theme of 'we demand great feasts of public space' (in Duncombe 2002, 220–1). This new style of dissent in New York soon distinguished itself from more traditional leftist actions (see Duncombe 2002, 221–3). The prime distinction was between 'proactive spectacle' and 'passive spectatorship'. In the latter, a centralized leadership organized a 'standard' demonstration with the customary placards and chanting. The objective was to garner sufficient numbers and colour to attract the attention of both their political targets and the media. By contrast, proactive spectacle embraced celebration rather than demonstration, launching a captivating politics of the carnival.

New York's experience of gentrification illustrated well the impacts of neo-liberal globalization on community wellbeing, and became a symbol of the struggle over public space and community autonomy. Activists throughout the globe carefully observed, and took heart, from this one city's resistance. While these kinds of activities were at their pinnacle in the late 1990s, RTS remains a tactical fixture in the contemporary repertoire of oppositional politics. It is now particularly active in anti-war actions that since the advent of the Iraq war are a more regular oppositional theme. But many events continue to retain their urban space reclamation themes. In Sydney, Australia a January 2005 party was situated in the same location at its original 1997 action. A local event to protest against the local council's 'heavy-handed management approach', it recommitted to 'car-free streets, better public transport, improved conditions for pedestrians, cyclists and skaters, more urban leisure space, street gardening, liveable neighbourhoods and re-humanising our streets' (RTS Australia n.d.). It reiterated its support for the reclamation of space from the enclosures of traffic, privatization and public land sales, so that community and freedom could prosper. It also reaffirmed carnival as its politics of pleasure and the strategic debt it owes to Situationism.

Carnival and the politics of pleasure

RTS draws inspiration from a number of historical sources, especially carnivals of rebellion such as the 1871 Paris Commune, the Suffragette parades and Paris 1968. The Situationist character of Paris 1968 and the carnivalesque style of dissent that it launched was to particularly enthuse them. RTS borrowed the ethics and tactics of their historical comrades, and adapted them into their effective contemporary forms. RTS finds in the oppositional repertoire of carnival a compelling strategy for the doing of their own forms of dissent. Central to the carnival's oppositional ethos is its exhortation to participate rather than simply spectate, and to celebrate rather than simply demonstrate. In this way, participation is directly opposed to the mediated passivity and conformity of a commodified culture. Influenced by the thinking of Mikhail Bakhtin (1984), RTS' activists embrace carnival as the political theatre of autonomy:

Carnival does not know footlights, in the sense that it does not acknowledge any distinction between actors and spectators. Footlights would destroy a carnival, as the absence of footlights would destroy a theatrical performance. Carnival is not a spectacle seen by people; they live in it, and everyone participates because its very idea embraces all the people ... Carnival laughter is the laughter of all people ... it is universal in scope; it is directed at all and everyone, including the carnival's participants. The entire world is seen in its droll aspect, in its gay relativity ... [T]his laughter is ambivalent: it is gay, triumphant, and at the same time mocking, deriding. It asserts and denies, it buries and revives. Such is the laughter of carnival (Bakhtin in Notes from Nowhere 2003, 178).

RTS actions set out to create just such a theatre of dissent. The RTS street party is essentially the revival of a 'Bakhitinian dialogic carnival, a loud clamour of a polyphonic, open, tumultuous, subversive conversation' (Scheurich 1997, 90).

Importantly, a carnival's form also creates a participatory egalitarianism that dissolves social rankings, and thus hierarchy, and enables all classes to intermingle in the spirit of enjoyment that they share. Carnival creates a temporary alternative to established and anticipated patterns of behaviour. It approximates a temporary autonomous zone that permits interim release from the external discipline of an imposing state. This 'undisciplined' behaviour directly challenges the status quo and hence the state's authority as a whole. Through the suspension of 'normal' life and through the intermingling of all social stratas, carnival temporarily transcends the conventional power relations that underpin life in a capitalist society, helping to break down 'social distance' through the 'construction of new forms of interpersonal relations' (Morrow 1998, 158–9). Carnival juxtaposes conventional social practices. It replaces the gravity of social change with pleasure, rioting with dancing, and physical and psychological drabness with colour and festivity. Reflecting this spirit of carnival, and speaking for the anti-globalization movement, the Direct Action Network announced that they would 'make revolution irresistible' because 'capitalism is boring'.

Strong anarchist impulses permeate the strategy of carnival. Bakhtin subscribes an 'almost anarchistic belief in the power of carnival'; as with many other anarchists, 'he recognises the importance of the everyday conditions that form a cauldron out of which the various ideologies arise' (Burkitt 1998, 176–7). RTS too acknowledges the importance of the everyday, turning the carnival-like street party into a revolutionary moment that celebrates the 'here and now' as against the 'there and later'. In carnival RTS thus found 'a model of protest in which the action itself was symbolic of its demands' – a melding, once again, of means and ends (Duncombe 2002, 222). Anti-globalization actions became more than just demonstrations; they were now 'carnivals against capitalism' – or, at the very least, against neo-liberal globalization. Even if implicitly, the anti-globalization movement perceived that resistance was strengthened when it was celebrated.

To achieve these goals, carnivals of rebellion thus had to transcend the traditional protest form: a uniform demonstration, an organizational hierarchy, the long march, a chosen destination, selected speakers, police escorts and ritual chants. In its place RTS promotes a resistance model that showcases the principles of diversity, creativity and autonomy in action, principles well encapsulated in the carnival:

> Throughout history carnival has been a time for inverting the social order, where the village fool dresses as the king and the king waits on the pauper ... This inversion exposes the power structures and illuminates the processes of maintaining hierarchies – seen from a new angle, the foundations of authority are shaken up and flipped around. The unpredictability of carnival ... [its] spontaneity ... creates a new world by ... turning the present world upside down, if only for a moment (Notes from Nowhere 2003, 174–5).

While doing their own form of autonomous resistance, the Zapatistas too appreciated the politics of carnival. Subcommandante Marcos observed that today's revolution has to be conducted very differently in order to be effective. He condemned the traditional form of socialist revolution, with 'men and women stoically marching behind a red, waving flag towards a luminous future', and urged that revolution should instead 'become a kind of carnival' (in Notes from Nowhere 2003, 176).

Since its politics is participatory rather than spectator driven, success relies on individual and collective input. A regular participant in anti-globalization and RTS events comments on carnival's inclusive and anarchical character: carnival 'demands interaction and flexibility, face-to-face contact and collective decision-making'; this enables the development of a 'dynamic and direct democracy' which has 'no leaders, no spectators, no sidelines, only an entanglement of many players who do their own thing while feeling part of a greater whole' (in Notes from Nowhere 2003, 178). Once again, the spontaneity and playfulness unleashed by carnival, highlights market society's converse offerings: an ordered, cautious and carefully orchestrated life. While the RTS street party might only last a few hours, the transformational psychology it unleashes is considered life changing. There is seemingly no going back once autonomy is tasted.

Situationism, the spectacle and culture jamming

In embracing the transformational psychology of a politics of pleasure, RTS exhibits a neo-Situationist, anarchical politics. Railing against the commodification and stifling of creativity, Situationists promoted a politics of the imagination practiced through the creative arts. They wanted 'the imagination, not a group of men, to seize power, and poetry and art to be made by all ... To hell with work, to hell with boredom! Create and construct the eternal festival' (Marshall 1993, 550). A leading Situationist theorist, Guy Debord, defined a situation as 'a moment of life concretely and deliberately constructed by the collective organization of a unitary ambiance and a game of events' (in Smith 2002a, 33). The RTS temporary street event is a fitting Situationist moment. The creative spectacle it generates, represents a politically conscious moment of resistance:

> The Situationist goal is immediate participation in a passionate abundance of life, through the variation of fleeting moments resolutely arranged. The success of these movements can only be their passing effect. Situationists consider cultural activity, from the standpoint of totality, as an experimental method for constructing daily life (Debord 1997, 90).

To work as a transformative moment of resistance in a capitalist regime, the situation thus replaces the passivity of life with moments that are active and *alive*. According to the original Internationale Situationniste, since 'the individual is defined by his [sic] situation, he wants the power to create situations worthy of his desires'; this includes replacing 'existential passivity with the construction of moments of life' and 'doubt with playful affirmation' (in Rasmussen 2004, 383).

The Situationist International was formed in 1957 from a combination of Guy Debord's Lettrist International, existentialism and some emerging art influences such as surrealism. Dissolved in 1972, its legacy now lives on in some of the oppositional politics of the 21st century. Its eclectic sources include medieval heresy, utopianism,

Dada, surrealism, soul music, and more recently, punk rock (Smith 2002a, 32). Primarily *post*-Marxist, but still anti-capitalist, its revolutionary philosophy resonated with that of an autonomist Marxism that did not invest its faith in the proletariat, revolutionary parties or vanguards as the main revolutionary forces. Instead, the Situationists emphasized the liberation of desire, creativity and spontaneity from the depths that capitalism had entombed them. Only by resurrecting the impulse for social change as a heart-felt yearning or *desire* and then developing the tools for its realization, could capitalism be permanently disabled. Transformation depended on creating a 'counter-aesthetic' to capitalism. As counter-aesthetics, situations released the desire for autonomy – a necessary impetus for social change. Situationism holds that transformational impetus, or 'revolutionary moments' cannot be contrived. Rather they arise as 'distinct "events" in which otherwise suppressed desires, frustrations and creativity can break loose' constructing in turn 'the conditions in which "self-organisation" can become a reality' (Tormey 2004, 174).

Situationists challenged many prevailing left orthodoxies and questioned the constitution of transformational strategy. They argued that revolutionary politics needed to attack capitalism at the point of consumption as well as the point of production since it was at the point of consumption that alienation was at its 'deepest' and its contradictions 'sharpest' (Smith 2002a, 34). Like Lettrists, Situationists also championed the revolutionary potential of youth. This represented a significant departure from prevailing conventions that invested revolutionary potential in the seasoned working class. While they continued to allot the working class an important revolutionary role, the Situationists theorized that to be truly successful, revolution needed the input of other alienated social groups, particularly students and other youth subcultures. Paris 1968 showcased a youthful revolutionary charge, as did the new social movements. Young people have been central to oppositional politics for some time, and the anti-globalization movement is now no different. While RTS does not target young people *per se*, its cultural and political roots – particularly in the rave, punk and artistic communities – are a magnet for the many young activists who reside in these communities.

The 'rave parties' organized by the youthful rave scene sub-culture influenced RTS's oppositional shape considerably (Luckman 2001a). Raves were usually all night dance parties fuelled by electronic dance music and the availability of 'dance drugs'. They became sub-cultural entertainment events organized in opposition to the domination and commercialization of popular music and entertainment. Constructing themselves as not-for-profit anti-commercial events for the enactment of fun, raves were also free parties that charged no admission fees. Their reliance on donations precluded the rental of official premises in any case, and the raves were often held in outdoor venues or abandoned 'squatted' buildings. Early raves were thus autonomous, DIY events organized by small collectives that utilized a grassroots grapevine to spread news of their upcoming 'happening'. Paradoxically, music corporations soon spied the commercial potential of these events and of this sub-culture so that by the mid 1990s raves had become an increasingly commercial phenomenon. Nonetheless, the influence of the early rave scene on RTS is clear, particularly in the anti-establishment DIY philosophy that it championed.

Raoul Vaneigem, a colleague of Debord's, captures the Situationist ethic well in the title of his influential text, *The Revolution of Everyday Life* (2001) where he seeks no less than the total dissolution of hierarchy. He finds that the 'repressive unity' of power manifest in three main ways: through coercion, seduction and mediation. He thus locates the 'revolutionary moment' in the minutiae of everyday existence whereby the individual's reclamation of moments of creativity and pleasure become inherently revolutionary acts. Since the state wields and finesses these repressive powers over individuals, subjective resistance unites to become a collective act of rebellion. For Vaneigem, vanguards are not the way to change the world: 'the world of –isms' and the 'ideologies of freedom: liberalism, socialism and Bolshevism' merely highlight the 'bloodbaths' conducted in their name (2001, 23). Demonstrating his appeal to contemporary activists, he states: 'People who talk about revolution and class struggle without referring explicitly to everyday life, without understanding what is ... positive in the refusal of constraints – such people have a corpse in their mouth' (2001, 26).

In embracing pleasure, spontaneity and creativity, the social control exerted by a capitalist reality is thus challenged. The state's most potent social controls are wielded through the disciplinary practices of a consumerist capitalism. Rebellion requires the inversion of the practices of capitalist society since it is only by turning this realism against itself that its power to ordain life is diminished. Rather than being linear and rational – the very values championed by capitalism – social change needs to be chaotic and random. Only in a non-circumscribed chaos can desire be unleashed and reality remade. Transformational success relies on challenging capitalism both internally and externally. This releases a 'process by which capitalism would be *subverted from within* as a prelude to its displacement by other ways of living' (Tormey 2004, 53).

For many, the oppositional spirit of Paris 1968 had embraced this thinking and its underpinning ethos was thus Situationist rather than Marxist. The determination to invert a capitalist realism through the use of linguistic paradox generated some of the more famous slogans of the time: 'be realistic: demand the impossible'; 'it is forbidden to forbid'; and 'take your desires for reality'. Diverse and spontaneous, Paris 1968 mixed traditional forms of protest with personal acts of liberation. This represented a very visual enactment of the insight that revolution needed to be both internal and external if capitalism was to be transcended. While Paris 1968 can be seen as Situationism's prime, its ideas continued to be promoted through the communication sources of largely anarchist presses (Smith 2002a, 32). The advent of global anti-globalization stirred a relatively widespread revival of interest. Animated by these Situationist ideas, RTS borrowed or produced their own famous slogans. The London RTS website (1997) opens with the announcement that 'We will claim nothing, we will ask for nothing. We will take. We will occupy'. At various times RTS has proclaimed: 'Beneath the pavement, the beach' (borrowing from the Paris 1968 slogan 'Defend the Collective Imagination. Beneath the cobblestones, the beach'); 'Poetry is in the streets' and 'In a society that has abolished every kind of adventure, the only adventure that remains is to abolish the society!'

RTS actions are thus 'spectacles' of dissent set on undermining the dominant capitalist aesthetic. This dissent lives on in the related *detournement* practices of 'culture jamming', 'adbusting' and 'subvertising'. *Detournement* refers to the juxtapositioning of capitalist 'signs' in an effort to undermine or subvert them. It represents the Situationists' key 'dialectical' technique. Literally meaning to turn against itself, *detournement* was the counter-offensive proposed by Debord for undermining the spectacle that was capitalism itself. *Detournement* is in essence 'plagiaristic', since it uses both the source and image of the original work to create a new work (BarbelithWebzine 2005). For Debord, the spectacle is 'capitalism accumulated until it becomes an image' (in Smith 2002a, 33). This image then obscures the very power of capital behind a wall of commodified pleasures. Rather than targeting Marxist scarcity, the Situationists instead targeted 'abundance and the contradictions it entailed – especially boredom, which they saw as an ultramodern, artificially created method of social control' (Smith 2002a, 33). Rebellion lay in mounting counter-spectacles against the capitalist spectacle of commodity fetishisation. Nowadays these fetishisations are increasingly enacted through advertising, product placement and the materialist commodification of culture. Because successful selling is in itself the art of seduction through the activation of desire, resistance needs an equally seductive counter-desire. This required the extraction from the psychological enclosure of capitalism a space, situation or moment of 'non-alienated' pleasure. Through 'turning around' or *detouring* capitalism's spectacles, images and commodities, Situationists reclaim and recast them. These situations would need to be not only 'resistant to the spectacle's most cunning seductions' but would also need to act as 'gesture[s] towards some future transformation of society' that creates 'pockets of non-alienation ... in preparation for the total transformation achieved by the revolution' (Puchner 2004, 7).

Culture jamming is considered a particularly effective *detournement*. The shifting of an image, sign or message out of its context invests it with a potent oppositional meaning. A 'subadvertised' sign would be one that is 'defaced' so as to diminish or ridicule its original intent. According to Lasn, much lauded jamming 'visionary' and editor of Adbusters, these oppositional tactics are widespread:

> Early culture jammers put graffiti on walls, liberated billboards, operated pirate radio stations, rearranged products on supermarket shelves, hacked their way into corporate and government computers and pulled of daring media pranks, hoaxes and provocations. A new generation of 'jammers' is organising 'critical massing' rallies and 'reclaim the streets' parties, launching social marketing TV campaigns, coordinating global events like Buy Nothing Day and TV Turnoff Week, jamming G-7 economic summits, initiating legal actions to revoke the charters of dysfunctional corporations, and pioneering an even more potent array of cultural interventions (Lasn 1999, 217).

The new technologies extend culture jamming opportunities considerably, but a favourite activity remains the defacing of billboards in public places. The hijacking of these billboards represents its own form of space reclamation against a dominant capitalism. The culture jammer's main intent is to 'fight fire with fire' by drawing, often spectacular, attention to what they consider to be the saturation of image pollution in public spaces. The acme of culture jamming belongs to the *Adbusters* organization which has been pivotal in finessing its form and its practice. They describe themselves as:

... a loose global network of artists, writers, environmentalists, eco-logical economists, media-literacy teachers, reborn Lefties, ecofem-inists, downshifters, rabble-rousers, incorrigibles, malcontents and green entrepreneurs. We are idealists, anarchists, guerrilla tacticians, pranksters, neo-Luddites, poets, philosophers and punks. Our aim is to topple existing power structures and forge a major rethinking of the way we will live in the 21st century ... and the way meaning is produced in our society (Lasn 1999, 111–12).

Since billboards are more numerous in poorer neighbourhoods, and since society's more vulnerable are often powerless to resist these encroachments, culture jammers feel compelled to advocate on their behalf. The defacing of these billboards thus become important polit-ical acts, attempts not only to rid public space of capitalism's 'waste' but also as fights for social justice more broadly: 'since most residents can't afford to counter corporate messages by purchasing their own ads, they should have the right to talk back to images they never asked to see' (Klein 2001, 310). Regardless of the geography of billboards, the fact remains that the streets in which they appear are public spaces 'polluted' by increasing capitalist signage and product placement. Culture jamming thus represents one more weapon in the tactical armoury of space reclamation. While it has long historical roots dating back to 1930s (Klein 2001, 335–40), culture jamming has been widely embraced as an effective, and enjoyable, resistance tactic. Like many of the tactics used by RTS, these ideas also demonstrate its anarchist temperament.

RTS and the anarchist temperament

RTS offers a philosophy and practice of dissent that straddles both a politics of pleasure and a politics of space. It not only celebrates the temporary liberation of public space from the physical and cultural enclosures of capitalism, but also promotes a DIY participatory ethos that links the means of a public space politics with the ends of a free society. Its strong anarchist impulses have helped shape its identity and its politics and it wears well the garb of post-ideological anar-chism. In both its theory and practice its politics of space endorses the broad contours of Bey's Temporary Autonomous Zone. Bey's self-described hippy/punk anarchism has exerted considerable sway on the form of oppositional politics embraced by RTS. This is not the same as saying that RTSers are Beyian devotees; the chances are that most have not heard of him. Rather, RTS and TAZ share the conviction that revolution must be both enjoyed and 'tasted'. In agreeing with Bey that 'one cannot struggle for what one does not know', RTS seeks to ensure that the autonomous practices of their street parties mimic the future they struggle for. In short, RTS accepts that an uprising must be *for* as well as *against* something. It is here that Bey's TAZ most strongly echoes the RTS party:

> The TAZ [or RTS action] is like an uprising that does not engage directly with the State, a guerrilla operation which liberates an area (of land, of time, of imagination) and then dissolves itself to reform elsewhere/elsewhen, *before* the State can crush it. Because the State is concerned primarily with Simulation rather than substance, the TAZ [or RTS action] can 'occupy' these areas clandestinely and carry on its festal purposes for quite a while in relative peace (Bey 2003, 99).

This is in any case a Situationist insight – a tradition Bey acknowledges he too draws from. He admits that his notion of liberating desire and the carving out of temporary free spaces represents the carrying to 'the next stage' the struggle begun by Situationism in 1968 and Autonomia in the 1970s' (Bey 2003, 62). RTSers are thus strong approximations of Bey's 'guerrilla ontologists', but Bey's guerrillas are also neo-Situationists in disguise.

Bey invokes Stephen Pearl Andrew's image of the anarchist society as a *dinner party* 'in which all structure of authority dissolves in conviviality and celebration' (in Bey 2003, 102–3). For Pearl Andrews the dinner party symbolizes 'the seed of the new society taking shape in the shell of the old' (in Bey 2003, 104). It is a place where individuality is fully 'admitted', interaction is free and participants spontaneously form and reform into smaller conversational subgroups before rejoining the social whole. Mutual respect and enjoyment pervades. Parties – dinner or otherwise – are thus spontaneous, participative and celebratory events that temporarily break down social stratas in the sharing of pleasures and conviviality. For both Pearl Andrews and Bey they are:

> ... face-to-face, a group of humans synergiz[ing] their efforts to realise mutual desires, whether for good food and cheer, dance, conversation, the arts of life ... in short, a 'union of egoists' (as Stirner put it) in its simplest form – or else, in Kropotkin's terms, a basic biological drive to 'mutual aid' (Bey 2003, 104).

In employing this conception of the dinner party, the RTS struggle for the right to party becomes a decidedly radical act. This party also combines the individualist and social strands of anarchism by providing a forum where individual participants (Stirner's 'union of egoists') combine in a spirit of 'mutual aid' to undertake a social project.

RTS also summon their anarchist impulses more directly. London's RTS uses a Bookchin insight as its central logo: 'Ultimately it is in the streets that power must be dissolved: for the streets where daily life is endured, suffered and eroded, and where power is confronted and fought, must be turned into the domain where daily life is enjoyed, created and nourished'. Demonstrating its anarchical goals, RTS describes itself as 'a direct action network for global and local social-ecological revolution(s) to transcend hierarchical and authoritarian society, (capitalism included)' (London RTS n.d.). Mirroring Bookchinite views, RTS charges that a liberated urban space provides abundant opportunity for an 'authentic politics'. In the creation of, albeit very temporary, self-managed communities of the street, the street party 'in theory, suggests a dissolution of centralised power structures in favour of a network of self-controlled localities' (London RTS 1997, 5). The Street Party thus becomes a community meeting where collectivity and autonomy are practised. As a body politic, the street party is in effect a commune that is (temporarily) self-governing, face-to-face and collaborative. Various elements of the street party are individually organized through different affinity groups. These individual affinity groups are finally cohered in the street party 'federation' and in their shared philosophy of the streets. For London RTS this represents a 'Network of networks, or, more appropriately: the Street Party of all street parties'; one that undermines 'centralised state and government structures' and constitutes a 'dual power' in direct opposition to them (London RTS 1997, 5). In whetting the appetite for social change, RTS hopes that these individual moments, and individual parties, will one day culminate in the Street Party of street parties – in short, in radical social upheaval:

> The ultimate street party ... is one where each person in each street in every village, town and city, joins with every other in rejecting capitalism, its exploitation and divisions. Indeed rejecting all hierarchy an domination, embracing instead an ecological vision of mutual aid, freedom, complementarity and interdependence. When the streets are the authentic social sphere for a participatory politics based on self-activity and direct action. When cooperation and solidarity are the social practice of society (London RTS 1997, 6).

Through the example of RTS, the philosophy of the streets and the politics of space have penetrated the practice of anti-globalization protest. As noted in Chapter 2, while mass anti-globalization events incorporate well worn protest strategies, there are also important differences. More so than their protest predecessors, contemporary anti-globalization protest is intent on the physical and symbolic occupation of public space to transform it into a living embodiment of the social alternatives envisaged (see Duncombe 2002, 228). In this way, RTS makes the street party the 'revolutionary moment' by confronting 'what is' with 'what could be', and in their enjoyment of a liberated moment, set out 'to re-energize the possibility of radical change' (London RTS 1997, 5). But in contributing its tactics to the broader AGM community, RTS recognizes that its politics also need to go further, especially if it is to realize the social change it champions. Its connection with networks such as the People's Global Action (PGA) helps it compensate for some of its political shortcomings. In attempts to strengthen dissent, PGA seeks to synthesize two important but 'discordant elements': the party and the politics. They do this by ensuring that the main political message of the party – the fight for global justice – are not obscured, and that these party-goers have clear oppositional targets in their sights: those political and corporate players that would destroy global justice.

Conclusion

RTS's clearly visible oppositional practices have helped define 21st century dissent. Its anarchical 'politics of space' highlight an important oppositional theme that we have already identified in our discussion of the World Social Forum in Chapter 2. This 'spatial' politics identifies the practice of autonomy in unmediated spaces outside the external control of the state, as pivotal to transformational strategy. RTS's street parties have added colour and enjoyment to oppositional politics even as the more serious business of changing the world remains. Through its embrace of a post-ideological anarchism that draws from a range of traditions and sources, RTS has assembled a compelling political philosophy and practice of the streets. Its style of dissent has in turn influenced the practices of anti-globalization protest as a whole.

Whether this street politics translates to effective transformational strategy is another matter altogether, however. The continued success of these strategies depends in part on deflecting the inevitable counter-offensives of the very system RTS places under siege. As we saw, raves and dance parties have increasingly become commercialized events with enormous profit potential. The culture jam can also work both ways. The jamming tactic has undoubtedly been successful in counter-appropriating signage from a dominant corporate culture. But this corporate culture has not necessarily been cowered by the assaults against it, launching instead its own tactical resistance. As Klein points out, rather than culture jamming representing a spanner in the works of corporate culture, 'marketers are increasingly deciding to join in the fun', recognizing that 'with its combination of hip-hop attitude, punk anti-authoritarianism and a well of visual gimmicks' the culture jam has 'great sales potential' (2001, 329). Smith (2002a, 34) too asks us to recall that the Situationist legacy from which RTS draws ended 'fragmented, isolated and defeated'. He goes on to warn that while Situationists may 'throw a great party' we should avoid 'fetishizing their failure' while 'romanticizing their integrity'. Debord was also concerned that the Situationists risked becoming the 'latest revolutionary spectacle' themselves, rather than those who cleverly employed its tactics (Puchner 2004, 13).

RTS has nonetheless utilized the spectacle of the street party very effectively. Many of these contemporary radicals would not in any case expect revolution to take place overnight. They recognize that consciousness change is a very protracted process and that the dynamic of assault and counter-assault underpins social change. Highly symbolic and highly visible, RTS still aspires to a very radical goal: the reclamation of public space from the enclosure of capitalism. It has revived a political strategy that goes a considerable way towards matching the means of social change with ends of a changed society. In situating autonomous direct action at the centre of its politics, RTS also proclaims its post-ideological anarchist heart.

8
Earth First!

Earth First! are direct action exemplars. They believe that it is only through their direct and resolute challenge to activities that degrade nature that they can effectively exercise their commitment to protect it. Earth First!'s name, and its consciously positioned exclamation mark, clearly identifies its objective: a steadfast and passionate defence of the earth. A member of the radical ecology movement, Earth First! draws directly from the deep ecology paradigm, has a strong wilderness focus and a decidedly anarchist temperament. While it does not necessarily specifically identify itself as anarchist, it incorporates strong anarchist impulses and its affiliates include many self-described anarchists. Claiming that it has no members as such, Earth First! characterizes itself more as a 'priority' than an organization, and its structure is essentially acephalous and affinity group based. It distinguishes itself from mainstream environmental organizations that it claims have been coopted by both the state and the corporate sector. Rather than continually negotiating, Earth First! responds directly and immediately when nature is threatened. Mistrustful of both (mainstream) environmental organizations and the organs of the state, Earth First! actions aim squarely at the corporate perpetrators of ecological damage. Utilizing an array of direct action strategies including civil disobedience, monkey wrenching and other acts of 'ecotage', Earth First! is determined to act rather than simply speak.

Earth First!'s heyday was the 1980s and 1990s but it continues to play a active role in the radical ecology movement today, as well as remaining a key component of the anti-globalization movement (AGM). This is despite its significant ideological conflicts with other radical ecologists in the past, especially with Bookchin's social ecology and many eco-socialists. Primarily focused on the protection of wilder-

ness against logging, dam building and destructive agricultural practices, Earth First! has more recently been involved in Reclaim the
Streets (RTS) and anti-roads activities (see Wall 1999; Smith 2000). As
noted in Chapter 7, Earth First! and Reclaim the Streets (RTS) are
closely linked organizations that share many principles, members and
action repertoires.

Through the anti-road movement, Earth First! helped instil in RTS its
animosity towards the car culture and influenced its embrace of direct
action. Both ideologically and tactically, the ensuing anti-roads movement inspired various RTS actions in which Earth First! often took part.
But Earth First! is first and foremost committed to radical ecology.
While sharing the principles and politics of various RTS and AGM
actions, its focus is primarily that of the earth's defender (see Manes
1990; Scarce 1990). Unapologetically earth-focused, its liberation rhetoric was very clearly not self- but other-directed, with nature the
significant other. Its allegiance to a 'militant' biocentrism sets Earth
First! apart from many other radical ecologists and radical politics as a
whole. While it shares with many other radical movements the desire
for autonomy and self-determination, it is an autonomy centred around
green values, and now increasingly some of the anti-civilizational values
of anarcho-primitivism.

From the outset Earth First! represented the sharp end of the nature/
culture debates that preoccupied radical ecology, both theoretically
and strategically, especially during its inception (Taylor 1991, 1994,
1999). But what distinguished Earth First! was not an ecocentrism that
many radical ecologists shared, but its militant biocentrism. The central principle of Earth First! is an unequivocal one: 'humans have no
divine right to subdue the Earth' since they 'are merely one of several
million forms of life on the planet'; they consequently 'reject even the
notion of benevolent stewardship as that implies dominance' and
believe that, mimicking Aldo Leopold, 'we should be plain citizens of
the Land community' (in Do or Die Editorial Collective 2003, 3). For
Earth First!ers, the planet, personified as Mother Earth, is a sacred
living being to whom human beings are 'ontologically interconnected'
(Fox 1990; Ingalsbee 1996, 268). To protect Mother Earth is merely an
act of protecting one's own family from the ravages of a marauding
invader.

Conceived this way, Earth First! defends itself through a number of
key claims (see Taylor 1995, 15–18). The first is the moral claim of deep
ecology as a whole: that nature has intrinsic worth, separate to any
instrumental value that humanity may relegate it. Second, Earth

First!ers articulate a very specific discourse of ecological crisis that acts to justify the taking of extreme measures. They claim that the planet's ecosystems are under unprecedented strain and that if left unchecked are in danger of collapsing altogether. In taking direct action to prevent such catastrophe, Earth First!ers are heroically protecting not only the earth but also a shortsighted humanity. The third claim is a political one: that the system of democracy is so flawed that even if one wished to pursue reform through institutional channels, a broken system would not permit it (Taylor 1995, 17). These three claims continue to underpin the political philosophy of Earth First! even as they are regularly reassembled in response to changing political climates. This chapter explores Earth First!'s drive to defend the earth in the context of these changing political climates. Considering Earth First!'s anarchical impulses throughout, the chapter traces the organization's evolution over the past three decades through to its participation in the AGM as a whole. It closes with an examination of a very radical green offshoot that is variously claimed or rejected as an Earth First! affiliate: the Earth Liberation Front.

The development of Earth First!

Earth First! draws from a long tradition of wilderness preservation in the United States. The reverence for wild nature that underlies the preservationist worldview is built on a respect for nature's integrity and magnificence. A precedent for preservationism was set long ago when a significant tract of wilderness was preserved as Yellowstone National Park in 1872, representing one of the world's first examples of nature preservation (Nash 1982, 108). Not all jumped onto the preservationist bandwagon however. The dispute between Gifford Pinchot of the United States Forest Service and John Muir of the Sierra Club exemplified well the continuing differences between a conservationist and preservationist approach to the environment, with the former seeking to '*conserve* nature *for* development', and the latter to '*preserve* nature *from* development' (Eckersley 1992, 39). This dispute also signified a shift from a 'Roosevelt-era conservationism, led by the patrician Save the Redwoods League, to the 1960s-era preservationism spearheaded by the Sierra Club' (London 1998, 156). Founded in 1892, the Sierra Club and the Save the Redwoods League at first shared much, especially during the first half of the 20ᵗʰ century. However, the accelerating felling of redwoods by the growing timber industries prompted the Sierra Club to adopt a more radical approach.

The Sierra Club's evolving biocentric values increasingly distinguished it from the more anthropocentric perspectives of conservationist organizations such as the Save the Redwood League. As London (1998, 160) points out, the Sierra Club and other sympathetic environmental organizations held an:

> ... ecocentric view [that] minimised humans' importance in the natural cosmos and emphasised the need to preserve non-human-influenced nature, that is, wilderness. No longer the champion of civilisation, redwoods were now cast as its victim. The ecocentric ethic exemplified by the 1960s-era redwood struggle is the basis of much of the modern environmental movement that locates itself as nature's defender against human arrogance and destruction.

Earth First! clearly demonstrated its credentials as a wilderness movement dedicated to the defence of the earth against its human despoilers. The earlier preservationists' struggles continued on in some of the more recent fights to save California's redwoods, of which Earth First! became an important part (see Schrepfer 1983; Harris 1995). In the discussion of deep ecology in Chapter 4, we noted that a wilderness preservation focus accommodates well the ethics of expanded moral standing on which biocentrism rests. Aldo Leopold's (1968) assertion that human beings are just 'plain members of the biotic community, not lord and master over all other species', strongly resonates the biocentrism of the preservationist paradigm. Preservationism nonetheless generated a spectrum of commitment from the moderate to the radical. Not all preservationists became radical Earth First!ers but those that did adopted from the outset a 'whatever it takes' attitude to protecting nature from accelerating onslaughts, even if this 'whatever it takes' did not include violence against people.

Earth First! (US) was originally conceptualized by several Arizona desert activists and disillusioned ecologists during a hiking trip in wild country. They included Dave Foreman, Howie Wolke, Mike Roselle, Christopher Manes and Bar Koehler. Formed around 1980, several figures, particularly Edward Abbey (but who was never a member) and Dave Foreman, quickly became Earth First!'s most controversial 'figureheads'. Dismissive of mainstream environmentalism, Dave Foreman announced that it was 'time for a warrior society to rise up out of the Earth and throw itself in front of the juggernaut of destruction' (in Do or Die Editorial Collective 2003, 1). In the first Earth First! publication in 1980, Foreman stridently asserted that, unlike the timid environmental movements before it, Earth First! would be politically uncompromising in its defence of the earth. The ensuing Earth First! logo became 'No Compromise in the Defence of Mother Earth'. Earth First!'s combativeness and steadfastness set it apart from not only the mainstream environmentalists they were contemptuous of, but also from the radical ecology movement as a whole. Their militant biocentrism increasingly distinguished, and continues to distinguish, them even from other radical ecology groups.

Another of Earth First! co-founders, Howie Wolke, articulated the Earth First! philosophy in a particularly forthright and combative manner. Proclaiming that 'from the vast sea of raging moderation, irresponsible compromise, knee-jerk rhetorical Sierra Club dogma and ... duplicity in the systematic destruction of the earth' there at least sprouts 'a small seed of sanity': Earth First! (Do or Die Editorial Collective 2003, 3). Furthermore, this 'small seed of sanity' would not resile from having to take necessary action to protect Mother Earth. This included violent action where necessary, but a property violence rationalized in terms of self-defence rather than provocation. As nature's advocates, Earth First!ers saw themselves as simply exercising a natural instinct towards self-defence in the face of forcible attacks on those under their care. For Wolke, non-violence was in any case 'unnatural' since 'the most basic animal instinct is to fight back when under attack' (in Nash 1990, 196). Moreover, violence in defence of a range of rights defined much of North American history; so that violence 'is as American ... as apple pie' (in Nash 1990, 196). Foreman defended and reiterated similar views. He too proclaimed that 'nearly all known organisms react with what can be called violence towards territorial encroachment and personal attack' and that ecologists protecting a defenceless nature were no different (in Nash 1990, 196). To pretend that nature could be defended otherwise was for Foreman at best shortsighted and at worst delusional. Although both Wolke and Foreman were referring to violence against property rather than people, these views nonetheless alarmed many fellow ecologists.

Earth First!'s early years (1980–86) coincided with the publication of their periodical Earth First! The Radical Environmental Journal, a forum for the promotion of its militant biocentrism. The fiction and non-fiction writings of Edward Abbey – a wilderness-focused environmentalist well before the formal entrance of the environment movement onto the global stage – was particularly influential during these early years. A United States writer and critic devoted to the protection of wilderness and public forests, Abbey experienced a similar epiphany to that of Aldo Leopold – whose *Sand Country Almanac* (1968) was instrumental in articulating and promoting ecocentrism. Abbey's work as a national park ranger helped forge insightful reflections on the plight of nature and humanity's relationship to it. These insights were recorded in his journals and eventually published as *Desert Solitaire* (1968) – an influential and moving eulogy to the grandeur of nature and the cruelty of the relentless assault upon it. While his views and 'solutions' were considered extreme by many, they touched a small but growing band of radical ecologists frustrated by the accelerating destruction of nature and the seeming nonchalance of many of their radical ecology colleagues in stemming it.

The publication of Abbey's (in)famous fictional piece, *The Monkey Wrench Gang* (1975), introduced the term monkey wrenching and the idea of eco-sabotage, or 'ecotage', into the lexicon and practice of radical environmentalism. While fictional, the novel was nonetheless roundly condemned by both the mainstream media and the mainstream environmental movement. The novel depicts the antics of a bunch of 'eco-warriors' who commit deliberate acts of sabotage against industrial developments detrimental to nature. While Abbey claimed the novel's intention was simply to entertain, it nonetheless inspired a small group of environmentalists to adopt and refine these new ecotage tactics. In the light of the failures of the more conventional strategies to arrest degradation, these more militant strategies offered considerable hope that a least some damage could be contained. Abbey's views were not always popular however, even among the growing numbers of radical ecologists of the day. But more distasteful than his ecology views, which drew a largely sympathetic response, were his reactionary political views. Dave Foreman was equally notorious for some of his controversial views, many of which he shared with Abbey. An encyclopaedic entry on Abbey notes that:

> Sometimes called the 'desert anarchist', Abbey was known to anger people of all political stripes (as well as environmentalists) ... [He] has been criticised by some for his comments on immigration and women. He differed from the stereotype of the 'environmentalist as politically-correct leftist', by disclaiming the counterculture and the 'trendy campus people' and ... by supporting some conservative causes such as immigration reduction and the National Rifle Association. He devoted one chapter in ... *Hayduke Lives* [1989] to poking fun at left-green leader Murray Bookchin (Wikipedia 2005).

Generally identified as an anarchist, Abbey produced a range of narratives that reconceptualized the good/evil dichotomy popular in mainstream storytelling. Abbey's fictional 'heroes' were rebelling not only against the destruction of Mother Earth but also from the ravages of an authoritarian culture determined to quash individual freedom and autonomy (Taylor 1995, 13). Nonetheless, Abbey was somewhat of a hero in the early years of Earth First! precisely because of, rather than despite, his notoriety and he spoke often at Earth First! rallies and assemblies.

Foreman's co-founders were not only concerned with the accelerating destruction of nature, but also with the seeming powerlessness of the environment movement to contain it. Their disenchantment with what they considered was the professionalization and hence de-radicalization of the environment movement, propelled them to forge a new approach and a new movement. On the basis of their disenchantment with environmentalism and guided by Abbey's innovative monkey wrenching tactics, Earth First! proceeded to promote, finesse and expand its profile. While at the editorial helm of the *Earth First! Journal* in the 1980s, Foreman collected an array of relevant articles and published them as the influential *Eco-defence: A Field Guide to Monkey-wrenching* (Foreman & Haywood 1987). This collection was essentially a 'how to' manual on the disabling (or monkeywrenching) of the many machines utilized to destroy wilderness. An assortment of ecotage tactics were promoted: the trashing of bulldozers, occupation of forests, subverting of billboards, digging up of logging roads, spiking of trees, invasion of offices, smashing of windows, disabling of animal traps and the scrapping of computers (Do or Die Editorial Collective 2003, 4).

From the outset, Earth First! (US) was uncompromising in its protection of wilderness against the ravages of development. Writing in 1982, Wolke unapologetically states:

> Earth First! has its roots deeply embedded in the wilderness of the western US. We've recognised that wilderness preservation is the most urgent necessity on Earth ... eventually, the gears of this civilisation will likely grind to a halt under the immense weight of its own blundering and greed ... So speak out with passion against the mindless insanity [of a range of environmental problems] ... But let Earth First! always concentrate its efforts on the wilderness battleground (1991, 247).

Winning this wilderness battleground required a forthright direct action plan. In the mid to late 1980s, Earth First! thus came into its own as a direct action exemplar. This was on the back of a number of novel and memorable campaigns that highlighted and honed its direct action repertoire. Earth First! was now well established as a radical ecology group committed to assertive and combative strategies in efforts to protect wild nature. Its innovative strategies to prevent logging, delay dam and bridge building and protect wildlife habitats, assured them plentiful media attention. Their notoriety spread not only among the general public but also within the by now well established environmental movement. Many environmentalists expressed as much distaste for Earth First! and its tactics as did state authorities and the media. This was no surprise to Earth First!, and simply reinforced their decision to dissociate themselves from an environmentalism that had capitulated to the demands of both corporations and the state.

The splitting of Earth First!

As determined as Earth First! was to resist assaults on nature, the state was equally determined to resist it. Earth First! found it hard to hold together in the face of increasingly negative publicity and the accelerating raids, arrests and court appearances that followed. In any case, all had not been harmonious within the branches of Earth First! for quite some time. A split had emerged in the late 1980s between supporters of Foreman and those of high-profile Californian activist, Judi Bari. Bari was critical of ecotage tactics, particularly that of tree-spiking which entailed the driving of large nails through tree trunks to prevent their felling. Bari's concerns were shared by a growing number of other Earth First!ers who were worried at the direction the movement's monkeywrenching focus was taking it. While the intent of tree spiking is to prevent logging and not threaten the safety of workers, an accident in 1987 saw a timber worker severely injured when his saw struck such a spike. While it was never clear who was responsible for the spiking, Earth First! was roundly condemned, a condemnation from which it never quite recovered.

Not only did Bari and her supporters condemn such tactics, they also appealed for cooperation between timber workers and the environment movement (see Bari & Kohl 1991). They were seeking, along eco-socialist lines, to inject a stronger social justice focus in Earth First!, one that considered the impact of environmental ruin on people as well as the planet. Seemingly innocuous, this move nonetheless represented a threat to Earth First!'s biocentric identity, and one that was vigorously resisted by many other Earth First!ers. Bari also challenged Earth First!'s 'old-boy network' and its 'cowboy politics'. Her passionate promotion of 'The Feminisation of Earth First!' in a 1992 essay of the same name sought the removal of gender biases that she claimed also riddled Earth First! (Wall 1999, 145; Shantz 2002). In proposing the formation of closer ties with workers, unionists and small business operators, Bari saw an opportunity for timber workers to join them in their fight against corporate power (see London 1998, 170). Bari was in turn roundly condemned by the Earth First! 'old guard' for such rapprochement with 'the enemy'. Foreman attacked her 'timid' ecologism, making it clear whose side he stood on. While he too expressed concern that someone got hurt in the tree spiking incident, he nonetheless reiterated that 'I quite honestly am more concerned about old-growth forests, spotted owls and wolverines and salmon' (in London 1998, 170). As a unionist as well as environmentalist, Bari had long voiced concern for the security of timber workers who became unwitting targets in the direct action campaigns of Earth First!. It was her green syndicalist views – views that included labour as an important component of the environmental community – that so inflamed the die-hard biocentrists in Earth First! and contributed to the ideological split that emerged (see Shantz 2004). Unrepentant, Bari defended her conception of Earth First!, contending that an Earth First! conceived in syndicalist terms is:

> ... not trying to overthrow capitalism for the benefit of the proletariat. In fact, the society we envision is not spoken to in any leftist theory that I've ever heard of. Those theories deal only with how to redistribute the spoils of exploiting the Earth to benefit a different class of humans. We need to build a society that is not based on the exploitation of the Earth at all – a society whose goal is to achieve a stable state with nature for the benefit of all species (1994, 57).

Her move to forge a stronger alliance with timber workers, and her criticisms of Earth First!'s direct action toolkit, incurred the wrath of many radical ecologists determined to put such traditional 'socialist' leftism behind them. The task of reconciling militant biocentrism with eco-syndicalism was looking increasingly impossible and the ensuing rift more permanent.

Bari is probably most renowned for the pipe bomb that exploded under her car seat in May 1990, and that injured her severely. She was then charged by state authorities as a Earth First! terrorist, and accused of having planted the bomb there herself in a terrorist act 'gone wrong', a mistake that turned perpetrator into victim. While Bari was cleared posthumously in 2002 and her estate compensated, at the time the US Federal Bureau of Investigation (FBI) and other state authorities were successful in implicating her and hence damaging Earth First!'s reputation (see Shantz 2003). As a result, Earth First! was considerably weakened, but not before it inspired the formation of other chapters across the globe, of which British Earth First!, as discussed below, was particularly successful.

By 1990 Dave Foreman and many of Earth First!'s earlier founders had in any case departed the movement. Foreman was arrested on a sabotage and conspiracy charge in 1990 but received a suspended sentence. He nonetheless ceased to be a spokesperson for Earth First! and co-founded a new organization, the Wildlands Project, in 1991, and an associated organization, the Rewilding Institute in 2003. He served on the Sierra Club's board of directors for several years in the 1990s but left when he was unable to garner support for his restricted immigration policies. These were views that had proved contentious when he was at the helm of Earth First!, and ones that he had clearly hung on to. From the start, Foreman had proved both an enigma and a contradiction. Along with Edward Abbey before him, Foreman had often presented a misanthropic and reactionary 'cowboy politics'. His views on population and those on people *vis-à-vis* nature were considered particularly offensive. With regard to the former he states: 'the only real hope for the continuation of diverse eco-systems on this planet is an enormous decline in human population'; if 'the AIDS virus didn't exist, radical environmentalists would have to invent one' (in Bradford 1987, 17). With regard to the latter, he charges: 'Call it fascist if you like but I am more interested in bears, rainforests and whales than in people' (in Bradford 1987, 18). Furthermore, 'the human race could go extinct, and I for one would not shed any tears' (in Tokar 1990, 14). Abbey before him also made a very forceful case for immigration restriction:

> In fact, the immigration issue is really a matter of 'we' versus 'they' or 'us' versus 'them'. What else can it be? There are many good reasons, any one sufficient, to call a halt to further immigration into the USA. One seldom mentioned, however, is cultural: ... If we allow our country – *our* country – to become Latinised ... we will be

forced to accept a more rigid class system ... less democracy ... a fear and hatred of the natural world, a densely over-populated land base, a less efficient and far more corrupt economy, and a greater reliance on crime and violence as normal instruments of social change (Abbey in Bradford 1987, 17).

As would be expected, these views drew a very critical response from other radicals and radical greens. A particularly bitter exchange was launched in *Fifth Estate* where Bradford (1987) charged that these extreme views were not only attributable to some misguided individuals. They also reflected the flawed logic of deep ecology as a whole – a logic that inevitably produced misanthropism. Bookchin too stepped into the critical fray, claiming that 'anything seems to pass through Deep Ecology's donut hole: anarchism at one extreme end and eco-fascism at the other', with the eco-fascism charge clearly directed at Earth First! (1988, 22). In the 1980s the antagonism between social and deep ecology was protracted and intense, with Earth First! significantly censured. An unexpected rapprochement between Foreman and Bookchin occurred when Foreman conceded that their continued conflict lost sight of the important values and goals that the radical ecology movement shared. In addition he admitted that his views may have been insensitive and analytically limited, acknowledging that he 'often left unstated, and sometimes unexamined, the social components of problems like over-population, poverty and famine, while trying to discuss their biological nature' (Bookchin & Foreman 1991, 107).

Despite such seeming accommodation many Earth First!ers remained defensive, not necessarily of reactionary political views, but of the overall political direction that Earth First! was taking. Laying their political cards clearly on the table, Foreman and Morton state:

In short, we see happening to the Earth First! movement what happened to the Greens in West Germany – an effort to transform an ecological group into a Leftist group. We also see a transformation to a more overtly counterculture/anti-establishment style, and the abandonment of biocentrism in favour of humanism. Mind you ... [w]e are generally supportive of such causes. But Earth First! has from the beginning been a wilderness preservation group, not a class-struggle group (1991, 264).

After Foreman and the 'old guard' left in the late 1980s, Earth First! underwent a considerable transformation, reinforcing Foreman's lament that the internal conflict over the 'heart' of the movement was always about its political identity.

The 'new' Earth First!, increasingly influenced by anarcho-primitivism, embraced a stronger anarchist identity while retaining its strong deep ecology roots. It also set about restructuring the organization along acephalous and decentralized lines. According to Earth First! Worldwide (n.d.):

> Earth First! is a priority, not an organisation. The only 'leaders' are those temporarily working the hardest and taking the most risks. New ideas, strategies and crucial initiative comes from individuals, and all decisions are made within affinity groups based on preferred tactics ... Our actions are tied to Deep Ecology, the spiritual and visceral recognition of the intrinsic, sacred value of every living thing.

Many of course claimed its strong anarchist impulses from the outset. Sierra Nevada Earth First! (n.d.) points out, for example, that:

> From the beginning Earth First! has been an anarchical movement. Really, Earth First! is a tribe existing in autonomous, consensus based groups who oppose the ignorance and destruction of industrial society and share a vision of a free, natural existence. No bureaucracy, no lobbyists, no organizational spokespersons, not even any membership. Earth First! happens when a group of committed activists decide together to stop further destruction of life.

Nonetheless, while unflinching in its continued commitment to protect the earth, the idea of enhancing autonomy and liberating the individual at the same time as liberating nature, came increasingly to the fore of the contemporary Earth First! rhetoric. Announcing that their direct actions only seem radical when compared to 'an entire paradigm of denial and control, where the individual is convinced they are powerless', their acts of rebellion were now more openly directed to the liberation of self as well as the liberation of nature (Earth First! Worldwide n.d.). A 'new generation' of Earth First! was equally determined to resist the encroachment of a 'civilization' that ruined all – people and nature – in its path.

Earth First! in Britain

Earth First! was launched in Britain in the 1990s, aligning itself and its campaigns to a wider anti-roads movement and associated RTS actions.

Attracting considerably less bad press than its sister organization in the United States, Earth First! (UK) was somewhat broader in its diagnosis of environmental ruin, accommodating social as well as the well rehearsed ecological concerns. One of Earth First!'s British co-founders, Jason Torrance, notes that:

> EF! in the States ... were coming from a biocentric equality for all life, deep ecology, point of view ... rather than the deep social-change issues. I still don't think Earth First! in the United States is in any way as social as it is over here ... In the States I think its' more coming from deeper ecological roots, seeing yourself as part of the world ... It really started as a wilderness preservation movement ... a radical movement to protect the last areas of wilderness and to reclaim some areas of wilderness (in Wall 1999, 145).

As we saw in Chapter 4, the wilderness issue is important in framing the national character of the green movement and the priorities that it identifies (see Hay & Haward 1988). Countries with significant tracts of wilderness or 'old growth' forests such as the United States and Australasia are more open to the ecocentric world view than those countries whose environmental activism focuses on pollution and other human-related impacts of degradation. Wilderness issues are often conceptualized as 'green' issues, raising protection of nature concerns; while pollution and other forms of degradation are conceptualized as 'brown', raising distributive concerns. A wilderness focus can thus condition the character of a country's environmental politics, with preservationism, as we saw, pivotal in shaping the US chapter of Earth First!

The fit, of course, is not always so neat, with the United States also developing a very strong environmental justice movement that focuses on the distributive impacts of environmental risk. Moreover, a paradoxical feature of the British Earth First! is its emergence as a wilderness defence movement with little wild nature to defend, even as it did incorporate a stronger social focus. Earth First! (UK) nonetheless adapted well the principle of defending nature against its corporate ravagers, and applied this principle successfully to the idea of anti-roads. It was also determined to distance itself from the reactionary politics of their US Earth First! cousins, seemingly agreeing with Bari that these views represented a wilful disregard for social justice. The close ties in Britain between the anti-roads movement and Earth First! emphasized the social as well as ecological impacts of, in this instance, the car culture. Much oppositional angst was directed at the accelerating destruction of nature demanded by the burgeoning number of roads and highways. But the associated social impacts of land and house reclamations, pollution and health effects and the disregard for community concerns were also afforded a prominent profile in the campaigns. British Earth First!'s close association with RTS also introduced a distinctive 'politics of space' that helped frame environmental problems in a way that linked their ecological and social impacts. As we saw in previous chapters, space is in many ways the very 'stuff of power', since it highlights the way the corporate world appropriates both culture and nature (Thrift 2000, 274). To resist this appropriation is thus a political struggle over the right to determine the utility of public space, with the outcomes of these struggles having significant ecological and social consequences.

Like its original US counterparts, Earth First! (UK) arose in response to the perceived weakness of Britain's mainstream environment movement. Much antagonism had been directed at one of Britain's most successful environment organizations, Friends of the Earth, for some time. A main concern was the increasing professionalization and deradicalization of such green organizations. This deradicalization was, paradoxically, a response to these organizations' success in penetrating mainstream political institutions, a not uncommon trajectory for many social movements. In response to the absorption of their concerns into the mainstream political agenda, many social movements have experienced a shift from grassroots activism to 'a process of deradicalisation, oligarchisation, institutionalisation and professionalisation' (Van der Heijden 1999, 201). Rootes (1995, 80) points out that because 'environmental campaigners in Britain ... enjoy relatively favourable access to decision makers' they are 'constrained against action which might compromise that access'. The broad impact can be the 'disciplining' of environmental action and of the influence green actors are able to exert over the policy agenda. The rise of Earth First! and other radical forms of green activism in Britain was a direct response to the limitations these institutional constraints imposed on the expression of green dissent.

The new green radicals in Britain began incorporating tactics borrowed from their radical ecology counterparts in both the US and Australasia. In particular, they imported the strategies of non-violent civil disobedience, media stunts and some of the tools of monkey wrenching (Do or Die Editorial Collective 2003, 6). These tactics were utilized in the first combined anti-road and RTS action at Twyford

Down in early 1992. According to Earth First! these tactics 'set a pattern of activism prevalent for much of the next decade – a cycle of
national actions, anti-road campaigns, office occupations, night-time
sabotage and street blockades' (Do or Die Editorial Collective 2003, 7).
Innovative and attention grabbing, these direct action tactics quickly
distinguished Earth First! and chapters of RTS from mainstream organizations like Friends of the Earth whose alleged comfort within mainstream institutions had well and truly extinguished any radical
impetus. Some observers went further, claiming a significant change in
the character of British environmentalism, led by this new generation
of direct action activists (Doherty 1999).

Doherty (1999, 275) identifies several factors in explaining the transformed character of British environmentalism. These include the
changing values of British youth in the 1990s, dissatisfaction with the
mainstream environment organizations, and the influence of radical
ideas and strategies on the shaping of new protest movements. But he
highlights 'counter-cultural ideas, ideologically-justified resistance to
formal organization and non-violent direct action' as central to the
spirit of this new dissent (1999, 276). As with many observers, he too
identifies the 1992 Twyford Down action as the first anti-roads protest
to showcase the new politics. As with many RTS actions, this action
was driven by two main agents. First, there was a diverse collection of
local groups increasingly disillusioned with mainstream protest ave-
nues and who were trying to more effectively protest against the negative impacts of road building on their local community. Second, the
local community was joined by a bevy of young, green radicals committed to a political philosophy of direct action as the only effective
means of social change (1999, 276). In short, Earth First! had clearly
influenced this new breed of British radical.

The 'new breed' set their sights not only on achieving an environmental outcome, as important as this was, but were equally determined to make their action an expression of community power.
Echoing their anarchical sensibilities, these new radicals fought for
both environmental, and individual and community integrity. They
were protecting not only nature against the enclosures of industrialism
but also the communities themselves against an encroaching conformity. As we saw in Chapter 7, the combination of RTS and Earth First! in
the anti-roads actions injected a decidedly anarchist impulse in the
fight for autonomy that these protests embraced. Cranking up their
activities in the mid to late 1990s, a period generally considered their
prime, the anti-roads actions attracted an increasing diversity of

activists and were successful in challenging the British Government's approach to roads policy generally. By 1997, with the demise of the national roads program, the anti-roads movement was claiming a triumphant victory.

During this time, the magazine *Green Anarchist* became one of the main meeting points for radical movements, signifying the more widespread embrace of anarchist ideas in the armoury of radical politics. Its pages contained lively debates on political philosophy, oppositional tactics and on the future of radical ecology and radicalism in general. Concerned by this new breed of radical and their unashamed attachment to anarchism, the state targeted *Green Anarchist* in the mid 1990s in a series of raids. Rather than containing it, the alleged heavy-handed tactics of the police instead raised *Green Anarchist's* profile and stimulated renewed support. According to Do or Die Editorial Collective (2003, 19) this 'repression' backfired so badly that 'an alliance of largely liberal publications swung behind them' with increased 'motions of support'. This included the support of green organizations that had kept their distance for some time, including the Green Party and Friends of the Earth.

With these victories behind them, Earth First! and the anti-roads movement foreclosed this chapter of their struggle and prepared themselves for the next. By 1997, two events generated a significant shift in the movement's focus (Do or Die Editorial Collective 2003, 23). First, its successes in the anti-roads movement meant that, paradoxically, with the decreasing number of land struggles, Earth First! lost its compelling activist beacon. However, an action against genetically modified (GM) agricultural products in Cambridgeshire in late 1997 provided Earth First! with its new activist 'hook'. Capitalizing on the growing concern over GM products, and on the direct action training and experience that they had accumulated in their anti-roads phase, the Cambridgeshire protest became the first of what would become over the next few years hundreds of similar, widely supported actions. This represented a new emphasis and a new direction for the radical ecology movement not only in Britain but throughout the globe.

The second development was the participation by many radical ecologists in global meetings of activists to discuss resistance against neo-liberal globalization. The meeting organized by the Zapatistas in the Mexican jungles in 1996 – the First Intercontinental *Encuentro* for Humanity and against Neo-liberalism – proved pivotal. These meetings helped fuse national struggles with a global resistance that consolidated in the AGM. Reflecting this global identity, Earth First! (UK) labels the period 1998–2002 their 'Consolidation and Global Resistance

Period', pre-empting their enthusiastic participation in the AGM of which they were and are an important 'green' part (Do or Die Editorial Collective 2003). The demise of the anti-roads campaigns prompted a considerable decrease in the number of Earth First! groups in Britain. But the anti-globalization movement provided a welcome impetus for a renewed engagement with radical politics. Even so, Earth First! (UK) contends that 'the radical ecological movement is in a surprising healthy state' and that, importantly, it has resisted the death knell of 'being assimilated into the mainstream' (Do or Die Editorial Collective 2003, 31). Earth First! is still 'active and still raw' and there remain many places that 'continue to be saved by ecological direct action' and many people that are 'still getting involved and inspired' (2003, 31).

The Earth Liberation Front

Over the past few years, however, a small number of those 'getting involved and inspired' are the extreme radicals of the largely United States based Earth Liberation Front (ELF) – a movement which has upped the ante considerably in their vigorous defence of the earth. The media attention they attract is out of proportion to their actual numbers, but their actions make for a considerable media cache. Often considered an offshoot of Earth First! – albeit a very radical one – many Earth First!ers, and particularly the post-ideological anarchists among them, still deny any links with it. ELF's association with the Animal Liberation Front, its labelling as a terrorist organization and its controversial and costly ecotage tactics, has ensured ELF a high media profile that eclipses that of Earth First! – and many other environmental organizations. In the United States ELF has claimed responsibility for millions of dollars worth of damage to car dealerships, particularly of Sports Utility Vehicles (SUV), corporate headquarters, new housing estates, holiday resorts, new road constructions and corporate machinery and tools. Its profile and activities have generated not only the label terrorist, but also its own category of terrorist organization: special-interest terrorism. In a statement to the US Senate Select Committee on Intelligence in 2001, special interest terrorists were described this way:

> Special interest terrorism differs from traditional right-wing and left-wing terrorism in that extremist special interest groups seek to resolve specific issues, rather than effect more widespread political change. Special interest extremists continue to conduct acts of polit-

ically motivated violence to force segments of society, including the general public, to change attitudes about issues considered important to their causes. These groups occupy the extreme fringes of ... political and social movements. Some special interest extremists – most notably within the animal rights and environmental movements – have turned increasingly towards vandalism and terrorist activity in attempts to further their causes (in O'Connor 2004).

The ELF has been most active in the United States where, in 2002, it was declared just such a domestic terrorist organization by the FBI. While its first acknowledged action is generally considered to have taken place in 1997, others trace its origins back to the Environmental Life Force founded in California twenty years prior. The duration of this original Environmental Life Force was shortlived, even if its logo lived on. This logo presented a confident elvan character dressed in suitably elvan clothes but with bullet straps draped over his upper body and armed with a large, but corked, shotgun. The next reclamation of the acronym as the Earth Liberation Front was allegedly by a British Earth First! group in Brighton in 1992, responding to the mainstreaming and deradicalization of Earth First! itself. Meanwhile, radical actions in the name of ELF began to take place in the United States from 1996 onwards. While origins and details are disputed or blurred, there is general agreement that by the late 1990s ELF was well established as an extreme radical environmental organization in both Britain and the US, but with more activity taking place in the latter.

The original Californian Environmental Life Force claims to have initiated its first and subsequent actions in California in 1977. The arrest of its founder John Hanna in late 1977 generated a period of self-reflection which culminated in the disbanding of the organization the following year (Hanna 2001). This dissolution was prompted by the negative publicity their guerrilla tactics were generating for the emerging environment movement. After laying low for many years, the Original ELF's founder agreed to an interview in 2001, alarmed at the ultra-radical turn his original namesake was taking. Initially prompted to 'go underground' in the defence of the earth because 'conventional methods of civil disobedience' had proved useless in arresting the degradation of nature, Hanna (2001) claims to now see that his behaviour was that of a 'frustrated sociopath'. Contending that it is impossible for ecologists to wear both hats of environmentalists and violent provocateurs since 'by definition an environmentalist cherishes all life', Hanna (2001) goes on to claim that:

Civil disobedience can be militant in practice and in perception. But violence is never civil. Thoreau and Ghandi landed in jail but they never could have accomplished their goals had they resorted to violence. They maintained the high moral ground and inspired support ... We don't need any more unabombers or idiots [like today's ELF] ... running around trying to change the world by coercion and intimidation.

Despite its description as 'eco-terrorist', the ELF's activities are directed against property rather than against life. At time of writing, ELF has not been responsible for any loss of life but they have inflicted significant property damage. Most direct action radical ecology activists are committed to non-violence – a position consistent with the eco-centric ethic to which many subscribe. However, many also distinguish between violence towards living beings and non-animate property, rationalizing the former as unacceptable, but the latter as justifiable if enacted in defence of the earth (Anderson 2004, 113). As we saw, ecotage and monkey wrenching have usually been defended in this way: as acceptable damage perpetrated against the *property*, or machinery, which damages the earth. As one activist contends:

> Monkey wrenching ... is direct action at its purest. If I stand in front of a digger then I may get hurt and the security guards may get hurt. But if I sneak out in the dead of night and decommission a piece of machinery, I'm not injuring anyone. I'm simply costing the company the maximum amount of money (in Anderson 2004, 113).

Generally speaking, however, Earth First!'s ecotage was minimal compared to that of ELF damage. The contentious issue thus becomes not so much damage to people, which all direct activism rejects, as the degree of property damage that ELFers are prepared to condone in the name of earth liberation. Consideration of these limits has polarized the environment movement, with the vast majority refusing to condone any. Many mainstream and radical environmentalists are alarmed that ELF's bad press has tarnished the important and peaceable work they themselves are doing on behalf of the planet.

ELF nonetheless interprets its charter to defend the earth literally and remorselessly. Its actions are directed primarily towards economic sabotage and it rationalizes that it has little alternative but to take even stronger measures against accelerating ecological damage. They consider that moderate tactics such as Earth First!'s monkeywrenching have proved futile in arresting this decline. In the face of this relentless assault, ELFers believe they are left no choice but to direct their sabotage where it hurts corporations most – financially. Traditional political measures and now an emasculated environment movement have spectacularly failed to save the earth. ELF sees governments as part of the problem rather than the solution and believe they often act as co-conspirators with the corporate sector in the devastation of nature. Craig Rosebraugh, a former ELF media spokesperson, captures well the zeal that drives many ELFers to the extremes of direct action, a zeal well encapsulated in the title of his 2004 book, *Burning Rage of A Dying Planet*. In a 2003 article in *Green Anarchist*, he implores ELF to:

> Think big. Wall Street, the stock market, Statue of Liberty, US Capital, Mt Rushmore ... large multinational corporations ...[This is the] difference between spraypainting and fire. When the ELF torched Vail Inc. resorts in 1998 the organisation was on the right path towards targeting desire (Rosebraugh 2003, 9).

Statements such as these are clearly provocative – and meant to be.

Many of these views are also reinforced in the more recent annals of the *Green Anarchist* journal. Its proud proclamation as a journal 'for the destruction of Civilization' and its warning that 'if you build it, we will burn it' do little to advance a moderate insignia, which is not its intent in any case. It is here that we find ELF's anarchist roots – albeit an ideological rather than post-ideological anarchism. ELF's anti-civilizational stand echoes parts of Zerzan's anarcho-primitivism, especially when it counts, as Zerzan does, the writings of the Unabomber Ted Kaczynski among its influences. *Green Anarchist's* continued dedication to 'the destruction of Civilization' and its regular publication of ELF articles reinforces these links. As expected with these blurred typologies, it is by no means clear whether anarcho-primitivists directly support the action of ELF, or whether ELFers are diehard primitivists. But ELF's organizational structure is clearly anarchic, subscribing to a decentralized, cellular and autonomous unit structure. Like Earth First!, it prides itself on the fact that it has no leaders, hierarchies or membership. It consciously conceptualizes itself as a 'front' and shares ideological and tactical similarities with other ideological anarchists such as the Black Bloc. As one supporter writes, 'as a "front" the ELF takes a bit of the Che Guevara image of third-world "national liberation" movements such as the Algerian "National Liberation Front" and extends it to a world scale' (Asan 2003, 11). In short, ELF has given new meaning to the word 'militant' in the descriptor militant biocentrism.

Conclusion

Earth First!'s direct action tactics have influenced the strategies of both the radical ecology and anti-globalization movements. With its philosophical roots in deep ecology and its political and tactical ones in anarchism, Earth First! constructs a militant biocentrism in the service of the planet. It recognized long ago that 'organizational hierarchy was the hallmark of the industrial monolith' that 'stifled wildness and diversity' (Devall in Anderson 2004, 114). Like most organizations Earth First! has undergone some significant changes during its time, but it has held on steadfastly to its commitment to protect the earth from further violation. As we saw, the Earth First! of twenty five years ago was nonetheless an appreciably different one.

Continuing to draw from the deep ecology, eco-anarchist and bio-regionalist roots of its forebears, today's Earth First! increasingly incorporates the insights of anarcho-primitivism and neo-Situationism. Commenting on a recent Earth First! rendezvous, a contemporary Earth First!er notes that one of the main aims of the 'new generation' of Earth First! is 'to add more anticivilization analysis and practice into the movement, linking it more firmly to a momentum against civilization itself' (Skunk 2005/6). This commentator also notes Earth First!'s considerable shift 'from its more right-of-centre wilderness conservationist beginnings in the '80s, which embraced sabotage and isolationism, to a liberal/ Leftist eco-social movement' (Skunk 2005/6). While acknowledging the contributions of its earlier model, another important objective is to 'infuse' more anarchist ideas into Earth First! and 'push the tactical envelope' (2005/6).

The organizational principles of Earth First! clearly testify to its anarchist impulses: 'membership' respects diversity rather than formal membership requirements; representative institutions are eschewed in preference for direct 'citizenship rights'; there is organizational spontaneity with 'few formalized decision procedures'; divisions of labour are limited, with role rotation paramount; leadership or elite positions are 'severely circumscribed in authority and tenure'; and organizational loyalty or 'maintenance' is not viewed as a goal or priority (Wall 1999, 152).

Earth First! today contains an eclectic mix of anarchists, deep ecologists, eco-socialists, Reclaim the Streeters, anti-roaders, primitivists, neo-Situationists and global justice activists. Its long standing anarchical temperament now tends to manifest in two main ways. On the one hand, Earth First! retains its anarchical underpinnings but is

increasingly the province of the more numerous post-ideological anarchists drawn to the force of many of its ideas and ideals, as well as its
heroic commitment to defend the earth. This 'new generation' also
tends to come and go, moving in and out of various oppositional
movements and specific protests as their priorities ordain. Our contemporary Earth First!er notes that while they 'hypothetically support a
wide range of tactics', in reality the new generation Earth First!ers 'only
employ and advocate for coalition building, educational outreach,
banner hands, lockdowns, tree-sits, and legal "monkeywrenching"'.
On the other hand, those smaller numbers dismissive of such seemingly insipid tactics look to Earth First!'s more extreme offshoot, the
Earth Liberation Front, which offers a more 'full-bodied' ideological
anarchism.

Conclusion: Towards 21st Century Dissent

This book has argued that anarchism has adapted itself well to its contemporary times. Seizing the opportunities offered it by a mixture of political, socio-economic, cultural and technological developments, an invigorated anarchism rode the politics of globalization to propel itself into the heart of 21st century dissent. This was largely achieved through the medium of anti-capitalist, anti-globalization and radical ecology movements. But this 'new' anarchism is a considerably reconfigured one and makes its presence felt in particular ways. The term post-ideological anarchism has been used to describe it. It was also noted that the fluid, flexible and inexact nature of post-ideological anarchism – features necessary to its identity – makes neat classification difficult. Nonetheless, in Part I we examined some of the theoretical, intellectual and political developments that helped revive this interest in anarchism. And in Part II, we used several case studies to illustrate the various ways it was expressed and enacted, as well as the tensions within it.

Post-ideological anarchism refers to the strong anarchical *temperament* that runs through radical politics today. It signals a more fluid and open embrace of anarchist ideas in the armoury of oppositional politics. While still borrowing from the insights of the old anarchist greats, the post-ideological anarchist impulse rejects the constraints of ideology, doctrinal purities and vanguardist politics. Instead post-ideological anarchists prefer to mix their anarchist principles with an assortment of other political ideas and traditions as they construct their own autonomous politics. An important new influence is environmentalism, particularly radical ecology, which most new anarchism has now embraced. Key anarchist principles remain of course; otherwise it would not be a post-ideological *anarchical* politics. Post-ideological anarchists are a diverse collection, from those who simply dip their toes in at the margins, such as the many young radicals in the anti-globalization movement, to those who construct a very specific politics in its wake, such as the Zapatistas. But all of these radicals are compelled by anarchism's staunchly autonomous, democratic and anti-hierarchical spirit.

There continue to be many 'diehard' ideological anarchists in radical politics, of course. While important and influential, they do not however necessarily exert the most inspiration on the tenor of contemporary radicalism today. The Black Bloc, for example, is often associated, in the general public's mind at least, with what anarchism is today. The Bloc are, after all, 'a collection of anarchists and anarchist affinity groups that organize together for a particular protest action' and who 'convey an anarchist critique of whatever is being protested that day' (Black Bloc n.d.). Highly decentralized and spontaneous they are best described as a 'tactic' than an organization, and have influenced the formation of Italy's *Ya Basta* and *Tute Bianche* (all white) – important players in the AGM and particularly in the Genoa protest (see Albertani 2002). The Black Bloc are undoubtedly important players in the politics of anti-globalization and as autonomous affinity-group extraordinaires they demonstrate well their passion for anarchism. They exemplify many of the characteristics that help inspire the contemporary anarchist temperament: acephalous, autonomous and decentralized network structures that challenge hierarchy. But it is because of their identity as 'hard, pure anarchists' (Albertani 2002, 585), that many post-ideological anarchists demur in embracing their form. Many find the combative precision of the Black Bloc's protest form uncomfortable. For post-ideological anarchists, while applauding the Black Bloc's resolute autonomy, there is some hesitation in accepting an oppositional politics that can resonate doctrinal and tactical prescription.

That said, this conclusion does not intend proceeding in the conventional manner of providing a summary of material covered. Rather it wants to draw on its findings to consider some broader conclusions about the character of contemporary dissent as a whole. It utilizes the recent works of Michael Hardt and Antonio Negri – works that resonate post-ideological anarchist impulses – to do this. A consideration of some of the issues they raise helps us to speculate on the character of radical politics in the early 21ˢᵗ century, and global politics as a whole.

A radical democracy for the 21st century

Michael Hardt and Antonio Negri have often been called anarchists, a label that they nonetheless reject. Most agree, however, that their recent writings have helped inspire contemporary dissent, albeit in some parts of the world more so than others. *Empire* (2000) is widely considered one of the key radical treatises of the new century; it has even been called the new Communist Manifesto 'for our times'. While an autonomist Marxist text, *Empire*'s analysis resonates anarchist impulses. Since autonomism has always been a close relative of anarchism, there is no surprise in this. It is precisely *because* of these anarchist impulses – often heard for the first time by many radicals not familiar with either Marxism or anarchism – that *Empire* has captured such widespread attention and support among today's radicals. Chances are that, as with many other seminal texts, anarchist or otherwise, most radicals have not read any of Hardt and Negri's works directly or in full.

Empire argues essentially that we live in radically different times, one that has produced a 'new form of sovereignty' (2000, xi). The old system of nation states has been superseded by a complex, supranational global network of power. But this new world is not like the imperialist one of old and is not directed by one superpowerful hegemon. Rather it 'establishes no territorial centre of power and does not rely on fixed boundaries or barriers'; it is instead a 'decentred and deterriorializing apparatus of rule that progressively incorporates the entire global realm with ... hybrid identities, flexible hierarchies, and plural exchanges through modulating networks of command' (2000, xi–xii). If this seems daunting, with *Multitude* (2004) they make clear that this 'new world order' presents oppositional politics with many transformative opportunities. And they identify the anti-globalization movement as just such a transformative and heterogenous politics of the multitude. Hope springs from their belief that 'contemporary capitalism, although seemingly impervious to anti-systemic challenge, is in fact vulnerable at all points to riot and rebellion' (Balakrishnan 2000, 144).

Autonomism has always embraced an anarchist given: that revolutionary potential lies beyond the working class. For anarchists, this potential is invested not only in those that Marx rejected – the dreaded lumpenproletariat and the peasantry – but also beyond class itself. As we saw, the dramatic break between anarchism and Marxism during the First Internationale was underpinned by these, and other, strategic considerations. Autonomism, as its name implies, promotes another anarchical insight: that participatory and *autonomous* forms of worker and community organizations are best able to transform society. It also invests labour with a more prominent role in the determination of capitalism's dynamic. But while continuing to invest a central role to the working class and the notion of class struggle, the constitution of this class is also considerably widened so that it includes a broader range of workers and other peoples. Hardt and Negri call this wider constituency the multitude. They distinguish their multitude from a 'unitary' and exclusionist conception of working class that ignores the globalization-induced transformation of work, a transformation that defines work not simply as industrial but also social production (2004, xv). Sounding decidedly anarchistic and Situationist, they claim that the multitude 'can never be reduced to a unity or single identity' since it is composed of 'different cultures, races, ethnicities, genders, and sexual orientations; different forms of labor; different ways of living; different views of the world; and different desires' (2004, xiv).

We find in Hardt and Negri's recent works some elements of post-ideological anarchism. This is especially so in its post-left recognition that revolutionary potential lies beyond vanguards and political parties; that decentralized, diverse and autonomous organization is key to social change; and radical democracy sits at the heart of radical politics today. In fact, many post-ideological anarchists 'sniff out' in autonomism a distaste for revolutionary dogma and a challenge to hierarchy and authoritarianism – even when wielded by a supposedly revolutionary left. But reflecting their continued embeddedness in Marxism, Hardt and Negri's conception of autonomy still remains too circumscribed for many anarchists. Negri's earlier works in particular still conceptualized autonomy in terms of autonomy from capital and other power relations that pivot around capital, rather than anarchism's autonomy from hierarchical relations more generally (Cuninghame 1999). In addition they still attributed a strategic role to some kind of 'intellectual elite' who would advise on best ways forward in the anti-capitalist struggle (Flood 2002).

But in their conception of biopower as 'a form of power that regulates social life from within' (Hardt & Negri 2000, 23), they begin identifying the broader practices of hierarchy and domination. There is no surprise in this given that *Empire* draws significantly from poststructuralism, particularly Foucault, Deleuze and Guattari. Hardt and Negri readily acknowledge the debt they owe these theorists for the notion of 'biopower' and beyond. This also helps explain much of the criticism levelled against them by the traditional left. Their displacement of imperialism as hegemony, their failure to provide a clear strategic guide, their marginalization of the working class's revolutionary role, and their embrace of a 'politics of difference' notion with multitude, were, and continue to be, considered particularly problematic (see Callinicos 2001).

However, where Hardt and Negri's work taps into the pulse of radical dissent, and echoes the conception of post-ideological anarchism, is in their identification and championing of the autonomy and radical democracy that drives today's oppositional politics. A deeply democratic thrust underpins contemporary dissent, an impulse that is picked up strongly by *Multitude*. Here the impulses of both post-ideological anarchism and *Multitude* owes much to Laclau and Mouffe's conception of radical democracy, a radical democracy demanded by the AGM as a whole. Many radicals today agree with Laclau and Mouffe that the task of the radical left is not so much to 'renounce liberal-democratic ideology' as such but to properly apply it; that is, to 'deepen and expand it in the direction of a radical and plural democracy' (1985, 176). As we saw in Chapter 1, anarchism draws from the liberal tradition, even if it renounces the hollowness of the liberal democracy that practices under its name. Many radicals, and post-ideological anarchists in particular, would agree with Laclau and Mouffe that the traditional left has been crippled by its attachment to classism, statism and economism, or productivism (1985, 177). Indeed post-leftism constructs its position in similar terms. For Laclau and Mouffe, rather than seeing domination and hierarchy as simply 'incarnated in the state', it is 'clear that civil society is also the seat of numerous oppressions' and hence also a site of 'antagonisms and democratic struggles' (1985, 179). Authentic social change hence lies in a radical, plural democracy that refuses relations of domination and hierarchy wherever they appear and that at the very least minimizes mediation.

Democracy sits at the core of this authentic social change. The Zapatistas echo this democratic impulse robustly and eloquently. Marcos observes that the 'network of resistance' today has no 'organising structure; it has no central head or decision maker; it has no central command or hierarchies. We are the network, all of us who resist' (1996c, 117). The Zapatistas have long recognized that democratic change is 'the only alternative to war'. Their pursuit of a deeply democratic project is one shared by most of the radical actors that we have identified and discussed, and one that pulsates through oppositional politics today. Those committed to this quest see tightly circumscribed mediated relations and hierarchical practices as inimical to democracy, and to the development of the autonomy on which 'true' democracy rests.

Hardt and Negri too claim that since 'no other path will provide a way out of the fear, insecurity, and domination that permeates our world at war', democracy has never 'been more necessary' (2004, xii). They identify, as we has done throughout, that globalization both threatens and provides opportunities for realizing justice and democracy. They identify two characteristics in particular that invests multitude with immense democratic potential. The first is the very kind of economy that globalization unleashes – one based on networked relationships that launch social production as 'the primary characteristic of the new dominant forms of labor today' (2004, xv). The second, and most important for our argument, is multitude's political characteristic: its capacity for autonomous, decentralized and collaborative forms of organization. It is here that the 'multitude is working through Empire to create an alternative global society' (2004, xvii). Their discussion of the anti-globalization action in Seattle echoes the argument we have made throughout this book about the anarchical character of 21st century dissent:

> The magic of Seattle was to show that these many grievances were not just a random, haphazard collection, a cacophony of different voices, but a chorus that spoke in common against the global system. This model is already suggested by the organizing techniques of the protestors: the various affinity groups come together or converge not to unite into one large centralized group; they remain different but independent but link together in a network character (2004, 288).

Closing remarks

This book has identified a strong anarchist temperament running through oppositional politics today. While it has presented this development in quite buoyant terms, it is not oblivious to the many serious problems dogging its politics. Some of these problems were noted in the chapters, even if they were not elaborated on at any length. Assessing effectiveness is in any case a very different task, and not the task we set ourselves in this book. But it is important to note that, at the time of writing, it is a resurgent 'traditional' socialism that appears to be capturing both government and world attention, especially in Latin America. The socialist Venezuelan president, Hugo Chavez has made a compelling case for a new kind of 'socialism for the 21st century', and it seems that many other socialist leaders in other Latin American countries are following suit.

Despite this, we have also detected another important, even if embryonic, current: a growing rapprochement between the 'best' and 'worst' of anarchism and socialism, and 'pieces' of other linked traditions such as poststructuralism. We see this in the quasi-socialist politics of a 'postmodern' Zapatismo, but a Zapatismo that eschews vanguards, champions autonomy, practices command-obeying and seeks the reclamation of dignity rather than state power. We see it in the arguments of proclaimed Marxists such as John Holloway who are talking of 'changing the world without taking power', and of substituting a 'counter-power' with an 'anti-power', the very form exercised by the Zapatistas (2002a, 36). Importantly, he identifies this 'anti-power' as the place where 'old distinctions between reform, revolution and anarchism no longer seem relevant, simply because the question of who controls the state is not the focus of attention' (2002a, 21). We see it in Hardt and Negri's recent autonomism that, while disavowing any anarchism, promotes the autonomous and heterogenous politics of the multitude. And we see it running through much of the radical politics that we have explored in our case studies and beyond. In discussing the contemporary 'soul of socialism', Blackman (2005) also notes its changing character. He asserts that there are signs that 21st century socialism is beginning to recognize that 'differences are OK' and that, like the Zapatistas, 'it is the basic right of any group to decide how they wish to be governed' (2005, 110). Importantly, he notes that the soul of socialism resides 'in the struggles of people who don't call themselves socialist' and in movements that 'are not always doctrinally orthodox' (2005, 110, 114).

A main attraction of this post-ideological anarchism for many radicals today is the freedom it offers for the autonomous determination of what is ideologically 'best' and the capacity to discard the rest. Once again, it is a politics that rejects the stranglehold of ideology and draws from a broader political canvas. This is not the same as saying that post-ideological anarchism is a political 'free for all': as we saw, it still contains key principles and key values, including decidedly green ones. But post-ideological politics in general, and a post-ideological anarchist politics in particular, do not proffer perfect models of dissent or singular visions of the good society. To do so would be to remain mired in an ideological cast that, instead of relying on autonomous visions of the good life, imposes social and tactical blueprints. The only 'fixed' political ideas to which post-ideological politics subscribe are the principle of autonomy and the practice of a non-hierarchical democratic politics.

One of the last words has to go to Neal (1997) who wrote before Seattle but anticipated the post-ideological anarchist impulses infusing contemporary dissent. We have already noted his distinction between small 'a' and capital A anarchism, the former denoting a less ideological strand than the latter, along with his attribution of 'the dismal state' of the anarchist movement in 1997 to the dominance of Anarchist ideologues with their 'elites, factions, cliques and cadres'. He invokes Chomsky to reinforce his point:

It's an odd feature of the anarchist tradition over the years that it seems to have often bred highly authoritarian personality types, who legislate what the Doctrine IS, and with various degrees of fury (often great) denounce those who depart from what they have declared to be the True Principles. Odd form of anarchism. (in Neal 1997).

We can surmise that Neal, and even Chomsky, would have been heartened to see the evolution of a more open post-ideological anarchist temperament, one that rejects 'roadmaps' of prescribed visions in favour of 'toolkits' for discovering them.

We reserve the very last word for the Zapatistas, however. They do not claim a blueprint for a new world, nor a revolution that will end up in a new class, faction or political group in power. Rather, the world they propose will 'end in a free and democratic space for political struggle'. This is a world where a 'network of voices ... resist the war Power wages on them'; and one 'in which sounds may be listened to separately, recognizing their specificity' but coming together 'in one great sound' (Marcos 2001, 46, 114). Inspiring, certainly; utopian, undoubtedly. But in chasing just such a dream, a post-ideological anarchist politics becomes a politics of and for its times.

Bibliography

Abbate, J. 1999. *Inventing the Internet*. MIT Press, Cambridge.

Abbey, E. 1968. *Desert Solitaire*. Simon and Schuster, New York.

Abbey, E. 1975. *The Monkey-Wrench Gang*. Avon, New York.

Adams, J. 2004. *Postanarchism in a nutshell*. Postanarchist Clearing House, Accessed 6th September 2004, http://www.spooncollective.org.

Adorno, T. 1973. *Negative Dialectics*. Seabury, New York.

Adorno, T. & Horkheimer, M. 1972. *Dialectic of Enlightenment*. Herber and Herber, New York.

Albertani, C. 2002. 'Paint it Black: Black Blocs, Tute Bianche and Zapatistas in the Anti-Globalisation Movement'. *New Political Science*, Vol. 24(4), 579–595.

Albrecht, G. 1993. 'Ethics and Directionality in Nature'. *Environmental Paradigms Conference*, University of New England, 16–18 April.

Anderson, J. 2004. 'Spatial Politics in Practice: the Style and Substance of Environmental Direct Action'. *Antipode*, Vol. 36(5), 106–125.

Andrews, S.P. 1852. *The Dinner Party*. The Science of Society, Accessed 26th October 2005, http://praxeology.net/SPA-DP.htm.

Arquilla, J. & Ronfeldt, D. 2001. *Networks and Netwars: The Future of Terror, Crime and Militancy*. National Defence Research Institute, California: RAND.

Asan 2003. 'The ELF and the Spectacle'. *Green Anarchist*, No. 68/9 Summer, 11.

Aviles, J. & Mina, G. 1998. *Marcos Y La Insurrecion Zapatista – La 'Revolucion Virtual' De Un Pueblo Oprimido*. Grijalbo, Mexico D.F.

Baker, G. 2003. 'Civil society that so perturbs: Zapatismo and the democracy of civil society'. *Space and Polity*, Vol. 7(3), 293–312.

Bakhtin, M. 1984. *Rabelais and his World*. Indiana University Press, Bloomington.

Balakrishnan, G. 2000. 'Virgilian Visions'. *New Left Review*, Vol. 5, 142–148.

BarbelithWebzine 2005. *Situationism in a Nutshell*. Accessed 4th October 2005, http://www.barbelith.com/cgi-bin/articles/00000011.shtml.

Bari, J. 1994. *Timber Wars*. Common Courage, Monroe.

Bari, J. & Kohl, J. 1991. 'Environmental Justice: Highlander after Myles'. *Social Policy*, Vol. 21(3), 71–77.

BBC News 2005. *A who's who of the G8 campaigners*. BBC News, Accessed 6th July 2005, http://newsvote.bbc.co.uk/mpapps/pagetools/print/news.bbc.co.uk/1/hi/scotland/4637609.stm.

Beck, U. 1994. 'The Reinvention of Politics: Towards a Theory of Reflexive Modernization'. In U. Beck, A. Giddens & S. Lash (eds), *Reflexive Modernization: politics, traditions and aesthetics in the modern social order*. Polity Press, Cambridge.

Beck, U. 1996. *Reinventing Politics: Rethinking Modernity in the Global Social Order*. Polity Press, Cambridge.

Beder, S. 1997. *Global Spin: The Corporate Assault on Environmentalism*. Scribe Publications, Melbourne.

Beder, S. 2001. 'Neoliberal Think Tanks and Free Market Environmentalism'. *Environmental Politics*, Vol. 10(2), 128–133.

Benjamin, M. 1995. 'Interview: Subcomandante Marcos'. In E. Katzenberger (ed.), *First World, Ha Ha Ha! The Zapatista Challenge*. City Lights Books, San Francisco.

Bey, H. 1991. *The Lemonade Ocean and Modern Times: A position paper*. Accessed 27th January 2006, http://spunk.org/library/writers/bey/sp000917.txt.

Bey, H. 1993. *Ontological Anarchy in a Nutshell*. Accessed 13th August 2005, http://deoxy.org/hakim/ontologicalanarchy.htm.

Bey, H. 2003. *T.A.Z. The Temporary Autonomous Zone, Ontological Anarchy, Poetic Terrorism*. 2nd edition, Autonomedia, New York.

Bimber, B. 1998. 'The Internet and political transformation: populism, community and accelerated pluralism'. *Polity*, Vol. 31(1), 133–149.

Bimber, B. 2001. 'Information and political engagement in America: the search for effects of information technology at the individual level'. *Political Research Quarterly*, Vol. 54(1), 53–67.

Black, B. 1986. *The Abolition of Work and Other Essays*. Inspiracy, Accessed 12th May 2005, http://inspiracy.com/black/.

Black, B. 1997. *Anarchy after Leftism*. C.A.L. Press, Columbia.

Black, B. 2004. 'Apes of Wrath'. *Green Anarchist*, Vol. 70, 6www.greenanarchist.org.

Black, B. n.d. *Theses on Anarchism After Postmodernism*. Accessed 30th November 2005, http://www.insurgentdesire.org.uk/theses.htm.

Black Bloc n.d. *What is the Black Bloc?* Accessed 4th December 2004, http://www.infoshop.org/blackbloc_faq.html.

Blackman, A. 2005. 'What is the Soul of Socialism?' *Monthly Review*, Vol. 57(3), 104–114.

Blixen, S. & Fazio, C. 1995. *Interview with Marcos about neoliberalism, the national State and democracy*. Accessed 22nd March 2005, http://flag.blackened.net/revolt/Mexico/ezln/inter_marcos_aut95.html.

Bohman, J. 2004. 'Expanding dialogue: the Internet, the public sphere and prospects for transnational democracy'. *Sociological Review*, Vol. 52(1), 131–155.

Bookchin, M. 1980. *Towards an Ecological Society*. Black Rose Books, Montreal.

Bookchin, M. 1987. *The Modern Crisis*. Black Rose Books, Montreal.

Bookchin, M. 1988. 'Social Ecology versus Deep Ecology'. *Socialist Review*, Vol. 18(3), 9–29.

Bookchin, M. 1990a. *The Philosophy of Social Ecology: Essays on Dialectical Naturalism*. Black Rose Books, Montreal.

Bookchin, M. 1990b. 'Radical Politics in an Era of Advanced Capitalism'. *Our Generation*, Vol. 21(2), 1–12.

Bookchin, M. 1991. *The Ecology of Freedom: The Emergence and Dissolution of Hierarchy*. Black Rose Books, Montreal.

Bookchin, M. 1992. *Urbanisation without Cities: The Rise and Decline of Citizenship*. Black Rose Books, Montreal.

Bookchin, M. 1994. *What is Communalism? The Democratic Dimension of Anarchism*. in Anarchy Archives, Accessed 1st November 2005, http://dwardmac.pitzer.edu//Anarchist_Archives/bookchin/CMMNL2.MCW.html.

Bookchin, M. 1996. *Social Anarchism or Lifestyle Anarchism: An Unbridgeable Chasm*. Spunk Library, Accessed 24th November 2003, http://www.spunk.org/library/writers/bookchin/sp001512/

Bookchin, M. 2003. 'The Communalist Project'. *Harbinger*, Vol. 3(1), 20–35.

Bookchin, M. & Foreman, D. 1991. *Defending the Earth*. Black Rose Books, Montreal.

Boyd, A. 2002. 'Irony, meme warfare, and the extreme costume ball'. In B. Shepard & R. Hayduk (eds), *From ACT UP to the WTO: Urban protest and community building in the era of globalization*. Verso, London.

Bradford, G. 1987. 'How Deep is Deep Ecology: A Challenge to Radical Environmentalism'. *Fifth Estate*, Vol. 22(3), 1–32.

Bryant, B. & Mohai, P. 1992. *Race and the Incidence of Environmental Hazards*. Westview Press, Boulder.

Bullard, R.D. 1990. *Dumping in Dixie: Race, Class and Environmental Quality*. Westview Press, Boulder.

Bullard, R.D. 1997. 'Environmental Justice: Challenges at Home and Abroad'. *Environmental Justice: Global Ethics for the 21st Century Conference*, Melbourne, Australia, 1–3 October.

Burbach, R. 2001. *Globalization And Postmodern Politics-From Zapatistas To High-Tech Robber Barons*. Pluto Press, London.

Burgmann, V. 2003. *Power, Profit and Protest: Australian social movements and globalisation*. Allen and Unwin, Crows Nest NSW.

Burkitt, I. 1998. 'The Death and Rebirth of the Author: the Bakhtin Circle and Bordieu on Individuality, Language and Revolution'. In M. Mayerfield Bell & M. Gardiner (eds), *Bakhtin and the Human Sciences: No Last Words*. Sage, London.

Buttel, F. 2003. 'Some reflections on the Anti-Globalisation Movement'. *Australian Journal of Social Issues*, Vol. 38(1), 95–116.

Call, L. 2003. *Postmodern Anarchism*. Lexington Books, Lanham.

Callinicos, A. 2001. 'Tony Negri in Perspective'. *International Socialism*, Vol. 92, 33–61.

Callinicos, A. 2004. 'The Future of the Anticapitalist Movement'. In H. Dee (ed.), *Anti-capitalism: Where Now?* Bookmark Publications, London.

Capra, F. 1983. *The Turning Point: Science, Society and the Rising Culture*. Flamingo, London.

Carson, R. 1962. *Silent Spring*. Penguin, Harmondsworth.

Carter, A. 1999. *A Radical Green Political Theory*. Routledge, London.

Castells, M. 1989. *The Informational City: Information Technology, Economic Restructuring, and the Urban Regional Process*. Basil Blackwell, Oxford.

Castells, M. 2001. *The Internet Galaxy: Reflections on the Internet, Business and Society*. Oxford University Press, Oxford.

Castells, M. 2004. *The Power of Identity*. 2nd edition, Blackwell, Oxford.

Chomsky, N. & Vodovnik, Z. 2004. *Anarchism Interview*. ZNet, Accessed 29th November 2005, http://www.zmag.org/content/print_article.cfm?itemID=6805&sectionID=41.

Cohen, R. & Rai, S.M. 2000. In R. Cohen & S.M. Rai (eds), *Global Social Movements*. Athlone Press, London.

Commission for Racial Justice 1987. *Toxic Wastes and Race in the United States*. United Church of Christ, New York.

Commoner, B. 1971. *The Closing Circle: Nature, Man and Technology*. Knopf, New York.

Couch, J. 2001. 'Imagining Zapatismo: the Anti-globalisation Movement and the Zapatistas'. *Communal/Plural*, Vol. 9(2), 243–260.

Cuninghame, P. 1999. *The Future at our Backs: Autonomia and Autonomous Social Movements in 1970s Italy*. Accessed 25th March 2005, http://www.iol.ie/~mazzoldi/toolsforchange/archive/papers/pap002.html.

Cuninghame, P. & Ballesteros Corona, C. 1998. 'A rainbow at midnight: Zapatistas and autonomy'. *Capital and Class*, Vol. 66, 12–22.

Cuomo, C. 1992. 'Unravelling the Problems in Ecofeminism'. *Environmental Ethics*, Vol. 14(4), 351–363.

Curran, G. 1999. 'Murray Bookchin and the Domination of Nature'. *Critical Review of International Social and Political Philosophy*, Vol. 2(2), 59–94.

Curran, G. 2001. 'The Third Way and Ecological Modernization'. *Contemporary Politics*, Vol. 7(1), 41–56.

Curran, G. 2004a. 'Anarchism, Environmentalism and Anti-globalisation'. *Interdisciplinary Environment Review*, Vol. 6(2), 22–32.

Curran, G. 2004b. 'Environment, Globalisation and Equality'. In K. Horton & H. Patapan (eds), *Globalisation and Equality*. Routledge, London.

Day, R. 2004. 'From Hegemony to Affinity: The Political Logic of the Newest Social Movements'. *Cultural Studies*, Vol. 18(5), 716–748.

Debord, G. 1997. 'Theses on the Cultural Revolution'. *October*, Vol. 79, 90–92.

Dee, H. 2004. 'Introduction'. In H. Dee (ed.), *Anti-capitalism: Where Now?* Bookmark Publications, London.

della Porta, D. & Kriesi, H. 1999. 'Social Movements in a Globalizing World: an Introduction'. In D. della Porta, H. Kriesi & D. Rucht (eds), *Social Movements in a Globalizing World*. Macmillan, Houndmills.

della Porta, D. & Mosca, L. 2005. 'Global-net for Global Movements? A Network of Networks for a Movement of Movements'. *Journal of Public Policy*, Vol. 25(1), 165–190.

Devall, B. & Sessions, G. 1985. *Deep Ecology: Living as if Nature Mattered*. Penegrine Smith, Salt Lake City.

DeWitt, R. 2000. 'Poststructuralist Anarchism: An Interview with Todd May'. *Perspectives on Anarchist Theory*, Vol. 4(2)online at http://www.geocities.com/ringfingers/mayint.html.

Diani, M. 2001. 'Social Movement Networks: Virtual and Real'. In F. Webster (ed.), *Culture and Politics in the Information Age: A New Politics?* Routledge, London.

Dick, S. 2005. 'Anarchists put chaos on agenda for summit'. *The Scotsman*, 10th March 2005, http://www.nadir.org/nadir/initiativ/agp/resistg8/media/0310 chaos.htm.

Do or Die Editorial Collective 2003. 'Down with the Empire! Up With the Spring'. *Do or Die*, Vol. 10, 1–101.

Dobson, A. 2000. *Green Political Thought*. 3rd edition, Routledge, London.

Dodson Gray, E. 1989. 'Nature as an Act of Imagination'. In S. Nicholson (ed.), *The Goddess Re-awakening*. Theosophical Publishing House, USA.

Doherty, B. 1999. 'Paving the Way: The Rise of Direct Action Against Road-Building and the Changing Character of British Environmentalism'. *Political Studies*, Vol. 47(2), 275–291.

Doyle, K. 1995. *Noam Chomsky on Anarchism, Marxism & Hope for the Future*. Red and Black Revolution, Accessed 29th November 2005, http://flag.blackened.net/revolt/rbr/noamrbr2.html.

Dryzek, J. 2005. *The Politics of the Earth: Environmental Discourses*. 2nd edition, Oxford University Press, Oxford.

Duncombe, S. 2002. 'Stepping off the sidewalk: Reclaim the Streets/NYC'. In B. Shepard & R. Hayduk (eds), *From ACT UP to the WTO: Urban protest and community building in the era of globalization*. Verso, London.

Earth First! Worldwide n.d. *About Earth First.* Accessed 14th October 2005, http://www.earthfirst.org/about.htm.

Eckersley, R. 1989. 'Divining Evolution: The Ecological Ethics of Murray Bookchin'. *Environmental Ethics*, Vol. 11, 99–116.

Eckersley, R. 1992. *Environmentalism and Political Theory: Towards an Ecocentric Approach.* UCL Press, New York.

Eckersley, R. 1999. 'The Discourse Ethic and the Problem of Representing Nature'. *Environmental Politics*, Vol. 8(2), 24–49.

Ehrlich, P. 1968. *The Population Bomb.* Ballantine, New York.

Eisler, R. 1990. 'The Gaia Tradition and the Partnership Future: An Ecofeminist Manifesto'. In I. Diamond & G.F. Orenstein (eds), *Reweaving the World: The Emergence of Ecofeminism.* Sierra Club Books, San Francisco.

Eltzbacher, P. 1908. *Anarchism: Exponents of the Anarchist Philosophy.* Libertarian Book Club, New York.

Epstein, B. 2001. 'Anarchism and the Anti-Globalization Movement'. *Monthly Review*, Vol. 53(4), 1–14.

Epstein, B. 2003. 'Notes on the Antiwar Movement'. *Monthly Review*, Vol. 55(3), 109–116.

Esteva, G. 2001. 'The Meaning and Scope of the Struggle for Autonomy'. *Latin American Perspectives*, Vol. 28(no 2 issue 117), 120–148.

Esteva, G. & Suri Prakash, M. 1998. *Grassroots Postmodernism: Remaking the Soil of Cultures.* Zed Books, London.

Flood, A. 2001. *Why are the Zapatistas different?* Chiapas Revealed, Accessed 5th August 2005, http://flag.blackened.net/revolt/mexico/pdf/revealed1.htm.

Flood, A. 2002. *Is the emperor wearing clothes? A review of Negri and Hardt's Empire from an anarchist perspective.* Accessed 25th March 2005, http://struggle.ws/pdf/empire.html.

Foreman, D. & Haywood, B. 1987. *Eco-defence: A Field Guide to Monkeywrenching.* 2nd edition, Ned Ludd Books, Tucson.

Foreman, D. & Morton, N. 1991. 'Good luck, Darlin'. It's been great'. In J. Davis (ed.), *The Earth First! Reader: Ten Years of Radical Environmentalism.* Peregrine Smith Books, Salt Lake City.

Foucault, M. 1969. *Discipline and Punish: The Birth of Prisons.* Tavistock, London.

Foucault, M. 1976. *History of Sexuality, Vol. 1: Will to Know.* Penguin, Harmondsworth.

Foucault, M. 1980. In C. Gordon (ed.), *Power/Knowledge: Selected Interviews and other Writings 1972–1977 by Michel Foucault.* Pantheon Books, New York.

Fox, W. 1990. *Towards a Transpersonal Ecology: Developing New Foundations for Environmentalism.* Shambala, London.

Free Software Foundation n.d. Free Software Foundation, Accessed 16th March 2005, www.fsf.org/.

Fromm, E. 1942. *The Fear of Freedom.* Routledge and Kegan Paul, London.

Fromm, E. 1976. *To Have or to Be.* Harper and Row, New York.

Fukuyama, F. 1992. *The End of History and the Last Man.* Hamish Hamilton, London.

Gallaher, C. & Froehling, O. 2002. 'New world warriors: 'nation' and 'state' in the politics of the Zapatista and US Patriot Movements'. *Social and Cultural Geography*, Vol. 3(1), 81–102.

George, S. 2004. 'Taking the Movement Forward'. In H. Dee (ed.), *Anti-capitalism: Where Now?* Bookmark Publications, London.

Gerlach, L. 2001. 'The Structure of Social Movements: Environmental Activism and its Opponents'. In J. Arquilla & D. Ronfeldt (eds), *Networks and Netwars: The Future of Terror, Crime and Militancy*. National Defence Research Institute, California: RAND.

Giddens, A. 2002. *Runaway World: How Globalization is Reshaping Our Lives*. Profile Books, London.

Glavin, M. 2004. 'Power, Subjectivity, Resistance: Three Works on Postmodern Anarchism'. *New Formulation*, Vol. 2(2)http://www.newformulation.org/4glavin.htm.

Goldman, B. 1992. *The Truth about Where You Live: An Atlas for Action on Toxins and Mortality*. Random House, New York.

Goldsmith, E., Allen, R., Alleby, M., Davoll, J. & Lawrence, S. 1972. *Blueprint for Survival*. Houghton Mifflin, Boston.

Gould, S.J. 1991. *Bully for Brontosaurus: More Reflections on Natural History*. Penguin, London.

Graeber, D. 2002. 'The New Anarchists'. *New Left Review*, Vol. 13(6), 61–73.

Green Anarchy Collective 2004. 'What is Green Anarchy? An Introduction to Anti-Civilization Anarchist Thought and Practice'. *Back to Basics*, Vol. 4(17), 35–41http://greenanarchy.org/index.php?action=viewwritingdetail&writingid=2834returnto=about.

Guerin, D. 1970. *Anarchism*. Monthly Review Press, New York.

Gutmann, M.C. 2002. *The Romance Of Democracy – Compliant Defiance In Contemporary Mexico*. University of California Press, Berkeley.

Habermas, J. 1970. *Towards a Rational Society*. Beacon, Boston.

Habermas, J. 1981. 'New Social Movements'. *Telos*, Vol. 49, 33–37.

Hajer, M. 1995. *The Politics of Environmental Discourse: Ecological Modernisation and the Policy Process*. Oxford University Press, Oxford.

Hammond, J.L. 2005. 'The World Social Forum and the Rise of Global Politics'. *Report on Social Movements*, Vol. March–April 2005, 30–39.

Hanna, J. 2001. *Interview with ELF Founder*. ELF (the Original), Accessed 11th November 2005, http://www.originalelf.org/interview_with_elf_founder.html.

Hansen, T. & Civil, E. 2001. 'Zapatista Timeline'. In J. Ponce de Leon (ed.), *Our Word is Our Weapon: Selected Writings Subcomandante Insurgente Marcos*. Serpent's Tail, London.

Hardin, G. 1968. 'The Tragedy of the Commons'. In R. Clarke (ed.), *Notes for the Future*. Thomas Hudson, London.

Hardin, G. 1977. 'Rewards of Pejorative Thinking'. In G. Hardin & J. Baden (eds), *Managing the Commons*. W.H. Freeman, San Francisco.

Hardt, M. & Negri, A. 2000. *Empire*. Harvard University Press, Cambridge.

Hardt, M. & Negri, A. 2004. *Multitude: War and Democracy in the Age of Empire*. Hamish Hamilton, London.

Hari, J. 2002. 'Whatever happened to No Logo?' *New Statesman*, 11 November 2002, 20–22.

Harris, D. 1995. *The Last Stand: The War between Wall Street and Main Street over California's Ancient Redwoods*. Time Books, New York.

Harvey, D. 1989. *The Condition of Post-Modernity: An Enquiry into the Origins of Cultural Change*. Basil Blackwell, Oxford.

Harvey, N. 2001. 'Globalisation and resistance in post-war Mexico: citizenship, difference and biodiversity conflicts in Chiapas'. *Third World Quarterly*, Vol. 22(6), 1045–1061.

Hay, P. 1988. 'Ecological Values and Western Political Traditions: From Anarchism to Fascism'. *Politics*, Vol. 8, 22–29.

Hay, P.R. & Haward, M.G. 1988. 'Comparative Green Politics: Beyond the European Context'. *Political Studies*, Vol. XXXVI.

Haythornthwaite, C. & Wellman, B. 2002. 'The Internet in Everyday Life: An Introduction'. In B. Wellman & C. Haythornthwaite (eds), *The Internet in Everyday Life* (pp. 2–41). Blackwell, Oxford.

Hegel, W.G.F. 1953. *Reason in History (R.S. Hartman translation)*. Bobbs-Merrill, Indianapolis.

Hegel, W.G.F. 1969. *Phenomenology of Mind (J.B. Baillie translation)*. Harper Touchbook, New York.

Hirsh, J. 1999. 'Who said we want a revolution?' *This Magazine*, Vol. 33(1), 18–22.

Hirst, P. & Thompson, G. 1996. *Globalisation in Question*. Polity Press, Cambridge.

Holloway, J. 2002a. *Change the World Without Taking Power*. Pluto Press, London.

Holloway, J. 2002b. 'Zapatismo and the Social Sciences'. *Capital and Class*, Vol. 78, 153–160.

Horkheimer, M. 1974. *Eclipse of Reason*. Seabury, New York.

Horton, D. 2004. 'Local Environmentalism and the Internet'. *Environmental Politics*, Vol. 13(4), 734–753.

Horton, K. & Patapan, H. 2004. *Globalisation and Equality*. Routledge, London.

Ingalsbee, T. 1996. 'Earth First! Activism: Ecological Postmodern Praxis in Radical Environmental Identities'. *Sociological Perspectives*, Vol. 39(2), 263–276.

Internet World Stats 2004. *Internet World Stats: Usage and Population Statistics, 2004, Internet Usage Statistics: The Big Picture*. The Internet Coaching Library, Accessed 8th January 2005, http://www.internetworldstats.com/stats.htm.

Johnston, J. 2000. 'Pedagogical guerrillas, armed democrats, and revolutionary counterpublics: examining paradox in the Zapatista uprising in Chiapas Mexico'. *Theory and Society*, Vol. 29(4), 463–505.

Jordan, T. 2002. *Activism! Direct Action, Hacktivism and the Future of Society*. Reaktion, London.

Katz, F. 1996. 'The Agrarian Policies and Ideas of the Revolutionary Mexican Factions Led by Emiliano Zapata, Pancho Villa, and Venustiano Carranza'. In L. Randall (ed.), *Reforming Mexico's Agrarian Reform*. M.E. Sharpe, Armonk, New York.

King, Y. 1990. 'Healing the Wounds: Feminism, Ecology and the Nature/Culture Dualism'. In I. Diamond & G.F. Orenstein (eds), *Reweaving the World: The Emergence of Ecofeminism*. Sierra Club Books, San Francisco.

Kingsnorth, P. 2003. *One No, Many Yeses: A Journey to the Heart of the Global Resistance Movement*. Free Press, Great Britain.

Kinna, R. 2005. *Anarchism*. OneWorld Publications, Oxford.

Klein, N. 2001. *No Logo*. Flamingo, Great Britain.

Klein, N. 2002. *Farewell to the 'End of History'*. Socialist Register, Accessed 7th March 2005, http://www.yorku.ca/socreg/.

Kropotkin 1975. In E. Capouya & K. Tompkins (eds), *The Essential Kropotkin*. Macmillan, London.

Kropotkin, P. 1939. *Mutual Aid*. Pelican, London.

Laclau, E. & Mouffe, C. 1985. *Hegemony and Socialist Strategy*. Verso, London.

Laclau, E. & Mouffe, C. 2001. *Hegemony and Socialist Strategy. Towards a Radical Democratic Politics*. 2nd edition, Verso, London.

Lasn, K. 1999. *Culture Jam: the Uncooling of America*. Eagle Brook, New York.

Leopold, A. 1968. *The Sand Country Almanac*. Oxford Press, New York.

London, J.K. 1998. 'Common Roots and Entangled Limbs: Earth First! and the Growth of Post-Wilderness Environmentalism on California's North Coast'. *Antipode*, Vol. 30(2), 155–176.

London Reclaim the Streets 1997. 'Reclaim the Streets'. *Do or Die*, Vol. 6, 1–10, Accessed 1st September 2005, http://www.eco-action.org/dod/no6/rts.htm.

London Reclaim the Streets n.d. *Beat the Bombers: Party for Peace!* Accessed 7th October 2005, http://rts.gn.apc.org/.

Lozano, F. 2005. *Zapatistas Issue Unwanted Memories for Mexicans*. Council on Hemispheric Affairs, Accessed 12th November 2005, http://www.scoop.co.nz/stories/WO0507/500109.htm.

Luckman, S. 2001a. *Practice Random Acts: Reclaiming the Streets of Australia*. Australian Broadcasting Commission, Four Corners 4 June, 2001, Accessed 16th September 2005, http://www.abc.net.au/4corners/dance/politics/04luckman.htm.

Luckman, S. 2001b. 'What are they raving on about? Temporary Autonomous Zones and "Reclaim the Streets"'. *Perfect Beat*, Vol. 5(2), 49–68.

Luke, T. 1988. 'The Dream of Deep Ecology'. *Telos*, Vol. 76, 65–92.

Malatesta, E. 1974. *Anarchy*. Freedom Press, London.

Manes, C. 1990. *Green Rage: Radical Environmentalism and the Unmaking of Civilization*. Little and Brown, Boston.

Marcos 1994a. 'Votan-Zapata or Five Hundred Years of History'. In J. Ponce de Leon (ed.), *Our Word is Our Weapon: Selected Writings Subcomandante Insurgente Marcos*. Serpent's Tail, London.

Marcos 1994b. 'War! First Declaration of the Lacandon Jungle'. In J. Ponce de Leon (ed.), *Our Word is Our Weapon: Selected Writings Subcomandante Insurgente Marcos*. Serpent's Tail, London.

Marcos 1996a. 'Civil Society That So Perturbs'. In J. Ponce de Leon (ed.), *Our Word is Our Weapon: Selected Writings Subcomandante Insurgente Marcos*. Serpent's Tail, London.

Marcos 1996b. 'Closing Words to the National Indigenous Forum'. In J. Ponce de Leon (ed.), *Our Word is Our Weapon: Selected Writings Subcomandante Insurgente Marcos*. Serpent's Tail, London.

Marcos 1996c. 'Second Declaration of La Realidad for Humanity and against Neoliberalism'. In J. Ponce de Leon (ed.), *Our Word is Our Weapon: Selected Writings Subcomandante Insurgente Marcos*. Serpent's Tail, London.

Marcos 1996d. 'Tomorrow Begins Today: Closing Remarks at the First Intercontinental Encuentro for Humanity and against Neoliberalism'. In J. Ponce de Leon (ed.), *Our Word is Our Weapon: Selected Writings Subcomandante Insurgente Marcos*. Serpent's Tail, London.

Marcos 2001. In J. Ponce de Leon (ed.), *Our Word is Our Weapon: Selected Writings Subcommandante Insugente Marcos*. Serpent's Tail, London.

Marcos 2004. *Two Flaws*. Accessed 14th September 2004, http://flag.blackened.net/revolt/mexico/ezln/2004/marcos/flawsAUG.html.

Marcos 2005. *Sixth Declaration of the Selva Lacandona*. ZNet, Accessed 5th July 2005, http://www.zmag.org/content/print_article.cfm?itemID=8218&sectionID=1.

Marcuse, H. 1964. *One Dimensional Man*. Beacon, Boston.

Marquez, G. & Pombo, R. 2001. 'Subcommandante Marcos – The Punch Card and the Hourglass'. *New Left Review*, Vol. 9, 69–79.

Marshall, P. 1993. *Demanding the Impossible: A History of Anarchism*. Fontana Press, London.

Mathews, F. 1991. *The Ecological Self*. Routledge, London.

Mathews, F. 1995. 'Community and the Ecological Self'. *Environmental Politics*, Vol. 4(4), 66–100.

May, T. 1994. *The Political Philosophy of Poststructuralist Anarchism*. Pennsylvania State University Press, University Park.

McQuinn, J. 2003. *Post-Left Anarchy: Leaving the Left Behind: Prologue*. Accessed 9th February 2006, http://www.infoshop.org/inews/article.php?story=03/11/12/7294780.

Meadows, D.H. Meadows, D.L., Randers, J. & Behrens, W. 1972. *The Limits to Growth: A Report for the Club of Rome's Project on the Predicament of Mankind*. Universe, New York.

Merchant, C. 1980. *The Death of Nature*. Macmillan, New York.

Merchant, C. 1987. 'Eco-feminism', *New Internationalist* pp. 18–20.

Merchant, C. 1992. *Radical Ecology: The Search for a Livable World*. Routledge, New York.

Millar, B. 2005. 'Irish anarchists sow seeds of flower power'. *Sunday Times*, 8 May, 11.

Miller, D. 1984. *Anarchism*. J.M. Dent and Sons, London.

Milstein, C. 2004. *Anarchism's Promise for Anti-Capitalist Resistance*. Institute for Social Ecology, Accessed 5th September 2005, http://www.social-ecology.org/article.php?story=20031028151308269.

Mitchell, L. 2005. 'Eco-village 'is model for us all''. *BBC News*, 5th July 2005, http://newsvote.bbc.co.uk/mpapps/pagetools/print/news.bbc.co.uk/1/hi/uk4654077.stm.

Mol, A. & Spaargaren, G. 2000. 'Ecological Modernisation Theory in Debate: A Review'. *Environmental Politics*, Vol. 9(1), 17–49.

Monbiot, G. 2000. 'No Way to Run a Revolution'. *The Guardian*, 10th May, Accessed online on 16th September 2005 at http://www.Monbiot.com/archives/2000/05/10/no-way-to-run-a-revolution/.

Moore 2003. In Notes from Nowhere (ed.), *We are everywhere: the irresistible rise of global anticapitalism*. Verso, London.

Moore, J. n.d. *A Primitivist Primer*. Accessed 12th January 2006, http://chromatism.net/primitivism/primer.htm.

Morland, D. 2004. 'Anti-capitalism and poststructural anarchism'. In J. Purkis & J. Bowen (eds), *Changing Anarchism: Anarchist Theory and Practice in a Global Age*. Manchester University Press, Manchester.

Morrow, R. 1998. 'Bakhtin and Mannheim: an Introductory Dialogue'. In M. Mayerfield Bell & M. Gardiner (eds), *Bakhtin and the Human Sciences: No Last Words*. Sage, London.

Mumford, L. 1991. 'Authoritarian and Democratic Technics'. In J. Zerzan & A. Carnes (eds), *Questioning Technology: Tool, Toy or Tyrant?* New Society Publishers, Philadelphia.

Naess, A. 1973. 'The Shallow and the Deep, Long-Range Ecology Movement: A Summary'. *Inquiry*, Vol. 16, 95–100.

Nash, R. 1982. *Wilderness and the American Mind.* Yale University Press, New Haven.

Nash, R. 1990. *The Rights of Nature.* Primavera Press, Leichardt NSW.

Naughton, J. 2000. *A Brief History of the Future: From Radio Days to Internet Years in a Lifetime.* The Overlook Press, Woodstock and New York.

Neal, D. 1997. *Anarchism: Ideology or Methodology.* Accessed 12th January 2006, http://www.spunk.org/library/intro/practice/sp001689.html.

Newman, S. 2001. *From Bakunin to Lacan: Antiauthoritarianism and the Dislocation of Power.* Lexington Books, Oxford.

Newman, S. 2002. 'Max Stirner and the Politics of Humanism'. *Contemporary Political Theory*, Vol. 1, 221–238.

Newman, S. 2003. *The Politics of Postanarchism.* Institute for Anarchist Studies, Accessed 22nd April 2005, www.anarchist-studies,org/article/article.

Notes from Nowhere 2003. *We are everywhere: the irresistible rise of global anticapitalism.* Verso, London.

O'Connor, J. 1987. *The Meaning of Crisis: A Theoretical Introduction.* Basil Blackwell, Oxford.

O'Connor, J. 1988. 'Capitalism, Nature and Socialism: A Theoretical Introduction'. *Capitalism, Nature, Socialism*, Vol. 1, 11–38.

O'Connor, J. 1994. 'Is Sustainable Capitalism Possible?' In J. O'Connor (ed.), *Is Capitalism Sustainable: Political Economy and the Politics of Ecology.* The Guildford Press, New York.

O'Connor, T. 2004. *Ecoterrorism.* Accessed 11th November 2005, http://faculty.ncwc.edu/toconnor/429/429lect16.htm.

Offe, C. 1985. In J. Keane (ed.), *Disorganized Capitalism: Contemporary Transformations of Work and Politics.* Polity Press, Cambridge.

Olesen, T. 2004. 'Globalising the Zapatistas: From Third World Solidarity to Global Solidarity?' *Third World Quarterly*, Vol. 25(1), 255–267.

Ophuls, W. 1977. *Ecology and the Politics of Scarcity: A Prologue to a Political Theory of the Steady State.* Freeman, San Francisco.

O'Riordan, T. 1976. *Environmentalism.* Pion Limited, London.

Panitch, L. 2002. 'Violence as a Tool of Order and Change: The War on Terrorism and the Antiglobalization Movement'. *Monthly Review*, Vol. 54(2), 12–32.

Parsons, H. 1977. *Marx and Engels on Ecology.* Greenwood, London.

Patomaki, H. & Teivainen, T. 2004. 'The World Social Forum: An Open Space or a Movement of Movements?' *Theory, Culture and Society*, Vol. 21(6), 145–154.

Pepper, D. 1993. *Eco-Socialism: From Deep Ecology to Social Justice.* Routledge, London.

Perlman, F. 1969. *The Reproduction of Daily Life.* Accessed 26th October 2005, http://www.spunk.org/library/writers/perlman/sp001702/repro.html.

Pickerill, J. 2003. *Cyberprotest: Environmental Activism On-line.* Manchester University Press, Manchester.

Plumwood, V. 1992a. 'Current Trends in Ecofeminism'. *The Ecologist*, Vol. 22(1), 10.

Plumwood, V. 1992b. 'Feminism and Ecofeminism: Beyond the Dualistic Assumptions of Women, Men and Nature'. *The Ecologist*, Vol. 22(1), 8–13.

Plumwood, V. 1993. *Feminism and the Mastery of Nature.* Routledge, London.

Proudhon, P.J. 1989. *The General Idea of the Revolution in the Nineteenth Century.* Pluto, London.

Puchner, M. 2004. 'Society of the Counter-Spectacle: Debord and the Theatre of the Situationists'. *Theatre Research International*, Vol. 29(1), 4–15.

Purkis, J. & Bowen, J. 1997. 'Introduction: The masks of anarchy'. In J. Purkis & J. Bowen (eds), *21st Century Anarchism: Unorthodox ideas for a new millenium*. Cassell, New York.

Purkis, J. & Bowen, J. 2004. *Changing Anarchism: Anarchist Theory and Practice in a Global Age*. Manchester University Press, Manchester.

Rasmussen, M.B. 2004. 'The Situationist International, Surrealism and the Difficult Fusion of Art and Politics'. *Oxford Art Journal*, Vol. 27(3), 365–387.

Raymond, E. 2000. *The Cathedral and the Bazaar*. Accessed 7th January 2006, http://www.catb.org/~esr/writings/cathedral-bazaar/cathedral-bazaar/index.html#catbmain.

Reclaim the Streets Australia n.d. *Saturday January 29 2005*. Accessed 12th October 2005, http://www.cat.org.au/rts.

Regan, T. 1983. *The Case for Animal Rights*. University of California Press, Berkeley.

Rich, P. 1997. 'NAFTA and Chiapas'. *The Annals of the American Academy of Political and Social Sciences*, Vol. 550, 72–85.

Ritter, A. 1980. *Anarchism: A Theoretical Analysis*. Cambridge University Press, Cambridge.

Rocker, R. 1938. *Anarcho-Syndicalism*. Phoenix Press, London.

Rootes, C. 1995. 'Britain: Greens in a Cold Climate'. In D. Richardson & C. Rootes (eds), *The Green Challenge: The Development of Green Parties in Europe*. Routledge, London.

Rosebraugh, C. 2003. 'Beyond the ELF'. *Green Anarchist*, No. 68/9 Summer, 9–10.

Ross, J. 1995. 'Who are they, what do they want?' In E. Katzenberger (ed.), *First World, Ha Ha Ha! The Zapatista Challenge*. City Lights, San Francisco.

Ross, J. 2003. 'The Zapatista at Ten'. *NACLA Report on the Americas*, Vol. 28(3), 11–15.

Ross, J. 2005. *La Otra Campana: The Zapatista Challenge in Mexico's Presidential Elections*. Counterpunch, Accessed 20th March 2006, http://www.counterpunch.org/ross11052005.html.

Sader, E. 2002. 'Beyond Civil Society: The Left after Porto Alegre'. *New Left Review*, Vol. 17, 87–99.

Sassen, S. 1997. 'Electronic Space and Power'. *Journal of Urban Technology*, Vol. 4(1), 1–17.

Sassen, S. 1998. *Globalization and its Discontents*. The New Press, New York.

Scarce, R. 1990. *Eco-Warriors: Understanding the Radical Environmental Movement*. Noble Press, Chicago.

Scheurich, J. 1997. *Research Methods in the Postmodern*. Falmer Press, London.

Scholte, J. 2000. *Globalisation: A Critical Introduction*. Macmillan, London.

Schrepfer, S. 1983. *The Fight to Save the Redwoods: A History of Environmental Reform, 1917–1978*. University of Wisconsin Press, Madison.

Shantz, J. 2002. 'Judi Bari and the Feminisation of Earth First!: The Convergence of Class, Gender and Radical Environmentalism'. *Feminist Review*, Vol. 70, 105–122.

Shantz, J. 2003. 'The Judi Bari and Darryl Cherney Lawsuit Against the FBI and Oakland Police: A Landmark Victory'. *Feminist Review*, Vol. 73, 166–171.

Shantz, J. 2004. 'Radical Ecology and Class Struggle: A Re-Consideration'. *Critical Sociology*, Vol. 30(3), 691–710.

Sheehan, S. 2003. *Anarchism*. Reaktion Books, London.

Shiva, V. 1988. *Staying Alive: Women, Ecology and Survival*. Zed Books, London.

Shiva, V. 1999. 'Ecological Balance in an era of Globalisation'. In N. Low (ed.), *Global Ethics and Environment*. Routledge, London.

Shor, F. 2006. *Counter and Anti Hegemony at the 2006 World Social Forum*. Zed Net/Activism, Accessed 26th February 2006, http://www.zmag.org/content/print_article.cfm?itemID=9661&sectionID=1.

Sierra Nevada Earth First! n.d. *History of Earth First!* Accessed 1st October 2005, http://www.sierranevadaearthfirst.org/main.asp?goto=history.asp.

Singer, P. 1975. *Animal Liberation: A New Ethics for Our Treatment of Animals*. Avon, New York.

Skunk, F. 2005/6. 'Report on the Earth First! Rendezvous'. *Green Anarchy*, No. 21 Winter, 74–75, online at http://www.greenanarchy.org/index.php?action=viewjournal&printIssueId=18.

Smith, G. 2000. 'Radical Activism'. *Environmental Politics*, Vol. 9(4), 150–153.

Smith, J. 2002a. 'Society of the Spectacle'. *The Nation*, Vol. 274(7), 32–34.

Smith, M. 2002b. 'The State of Nature: The Political Philosophy of Primitivism and the Culture of Contamination'. *Environmental Values*, Vol. 11(4), 407–425.

Soper, K. 1990. 'Feminism, Humanism and Postmodernism'. *Radical Philosophy*, Vol. 55, 11–17.

Spretnak, C. & Capra, F. 1985. *Green Politics: The Global Promise*. Paladin, London.

Stahre, U. 2004. 'City in Change: Globalization, Local Politics and Urban Movements in Contemporary Stockholm'. *International Journal of Urban and Regional Research*, Vol. 28(1), 68–85.

Starhawk 2003. *Webs of Power: Notes from the global uprising*. New Society Publishing.

Starr, A. 2000. *Naming the Enemy: Anti-corporate movements confront globalization*. Pluto, Australia.

Stirner, M. 1995. In D. Leopold (ed.), *The Ego and its Own*. Cambridge University Press, Cambridge.

Sugar, J. 1995. *Interview of Peter Lamborn Wilson*. Accessed 30th November 2005, http://www.hermetic.com/bey/pw-interview.html.

Taylor, B. 1991. 'The Religion and Politics of Earth First!' *Ecologist*, Vol. 21(6), 258–266.

Taylor, B. 1994. *Eco-Warriors and the Global Apocalypse: The International Emergence of Militant Environmentalism*. University of New York State Press, New York.

Taylor, B. 1995. 'Earth First! and Global Narratives of Popular Ecological Resistance'. In B. Taylor (ed.), *Ecological Resistance Movements: The Global Emergence of Radical and Popular Environmentalism*. State University of New York Press, Albany.

Taylor, B. 1999. 'Review Commentary: Green Apocalypticism: Understanding Disaster in the Radical Environmental Worldview'. *Society and Natural Resources*, Vol. 12, 377–386.

The Australian 2005. 'People's lives at stake: Geldof'. *The Australian*, 8th July 2005, 8.

Thrift, N. 2000. 'Entanglements of Power; Shadows?' In J. Sharp, P. Routledge, C. Philo & R. Paddison (eds), *Entanglements of Power: Geographies of Domination/Resistance*. Routledge, London.

Tokar, B. 1990. 'Eco-Apocalytics'. *New Internationalist*, August, 14–15.

Tormey, S. 2004. *Anti-capitalism: a beginner's guide*. OneWorld, Oxford.

Tormey, S. 2005. 'From Utopian Worlds to Utopian Spaces: Reflections on the Contemporary Radical Imaginary and the Social Forum Process'. *ephemera*, Vol. 5(2), 394–408.

Touraine, A. 1974. *The post-industrial society: tomorrow's social history: classes, conflicts and culture in the programmed society*. Wildwood House, London.

Truscello, M. 2003. 'The Architecture of Information: Open Source Software and Tactical Poststructuralist Anarchism'. *Postmodern Culture*, Vol. 13(3).

Tucker, R.C. 1978. *The Marx-Engels Reader*. W.W. Norton & Company, New York.

Van der Heijden, H.A. 1999. 'Environmental movements, ecological modernisation and political opportunity structures'. *Environmental Politics*, Vol. 8(1), 199–221.

Vanaik, A. 2004. 'Rendezvous at Mumbai'. *New Left Review*, Vol. 26, 53–65.

Vaneigem, R. 2001. *The Revolution of Everyday Life*. 2nd edition, Rebel Press, London.

Viejo 2003. In Notes from Nowhere (ed.), *We are everywhere: the irresistible rise of global anticapitalism*. Verso, London.

Wall, D. 1999. *Earth First! and the Anti-Roads Movement: Radical Environmentalism and Comparative Social Movements*. Routledge, London.

Wallerstein, I. 2002. 'New Revolts Against the System'. *New Left Review*, Vol. 18, 29–39.

Warren, K.J. 1990. 'The Power and Promise of Ecological Feminism'. *Environmental Ethics*, Vol. 12(2), 125–146.

Watson, I. 2002. 'An Examination of the Zapatista Army of National Liberation (EZLN) and New Political Participation'. *Democracy and Nature*, Vol. 8(1), 63–86.

Wayner, P. 2000. *Free for All: How Linux and the Free Software Movement Undercut the High-Tech Titans*. HarperCollins, New York.

Webster, F. 2001. *Culture and Politics in the Information Age: A New Politics?* Routledge, London.

Weis, T. 2003. 'Reinventing Revolution: The Continuing Struggle of the Zapatistas'. *Canadian Dimension*, Vol. 37(6), 14–16.

Weiss, L. 1998. *The Myth of the Powerless State: Governing the Economy in a Global Era*. Polity Press, London.

Weston, J. 1986. 'Introduction'. In J. Weston (ed.), *Red and Green: The New Politics of the Environment*. Pluto Press, London.

Westra, L. & Wenz, P.S. 1995. In L. Westra & P.S. Wenz (eds), *Faces of Environmental Racism: Confronting Issues of Global Justice*. Rowman & Littlefield, Lanham, MD.

White Jr, L. 1973. 'The Historical Roots of our Ecologic Crisis'. In I.G. Barbour (ed.), *Western Man and Environmental Ethics: Attitudes Towards Nature and Technology*. Addison-Wesley, London.

Wikipedia 2005. *Edward Abbey*. Wikipedia, Accessed 14th October 2005, http://en.wikipedia.org/wiki/Edward_Abbey.

Wissert, W. 2005. 'Zapatista Rebels Eye 2006 Mexico Election'. *Boston Globe*, Vol. (13th August 2005)www.boston.com/news/world/latinamerica/articles/2005/08/13/zapatista_rebels_eye_2006_election/.

Wolke, H. 1991. 'EF!'s Proper Role'. In J. Davis (ed.), *The Earth First! Reader: Ten Years of Radical Environmentalism*. Peregrine Smith Books, Salt Lake City.

WOMBLES News 2005. 'Carnival for Full Enjoyment – the full story'. *WOMBLES News*, 27th October 2005, http://www.wombles.org.uk/news/article_2005_07_19_3132.php.

Woodcock, G. 1963. *Anarchism*. Pelican Press, London.

World Social Forum 2002a. *What the World Social Forum is*. World Social Forum, Accessed 13th January 2006, http://www.forumsocialmundial.org.br/main.php?id_menu=19&cd_language=2.

World Social Forum 2002b. *World Social Forum Charter of Principles*. World Social Forum, Accessed 13th January 2006, http://www.forumsocialmundial.org.br/main.php?id_menu=4&cd_language=2.

World Social Forum 2004. *Who organizes it*. World Social Forum, Accessed 13th January 2006, http://www.forumsocialmundial.org.br/main.php?id_menu=3&cd_language=2.

Young, I.M. 1983. 'Feminism and Ecology, and Women and Life on Earth: Eco-feminism in the 80s – Book Review'. *Environmental Ethics*, Vol. 5(2), 175.

Zerzan, J. 1994. *Future Primitive*. Autonomedia, New York, Accessed 14th December 2005, http://www.primitivism.com/future-primitive.htm.

Zerzan, J. 2004. 'The Modern Anti-World'. *Green Anarchy*, Vol. 18, 16.

Index